NERVE AND MUSCLE EXCITATION

NERVE AND MUSCLE
EXCITATION

Douglas Junge
UNIVERSITY OF CALIFORNIA,
LOS ANGELES

SINAUER ASSOCIATES, INC.
SUNDERLAND, MASSACHUSETTS

FIGURE CREDITS

We gratefully acknowledge the cooperation of the following:

Chapter 6, Figure 3: Courtesy of Charles C. Thomas, Publisher, Springfield, Ill.

Chapter 6, Figure 4; Chapter 7, Figure 1; and Chapter 10, Figure 5: By copyright permission of The Rockefeller University Press.

Chapter 7, Figure 4: Copyright 1971 by the American Association for the Advancement of Science.

Chapter 7, Figure 6: Copyright 1971 by Litton Educational Publishing, Inc. Reprinted by permission of Van Nostrand Co.

Chapter 8, Figure 4: Copyright 1967 by the American Association for the Advancement of Science.

Chapter 9, Figure 2: Reprinted by permission of Pergamon Press.

Chapter 10, Figures 2 and 3: Used by permission of North-Holland Publishing Company, Amsterdam.

COVER

Variation of barnacle muscle action potential. From Hagiwara *et al.*, 1964. (See figure on page 104 and reference on page 111.)

NERVE AND MUSCLE EXCITATION

First Printing

Library of Congress Catalog Card Number: 75-30151

ISBN: 0-87893-408-1

FOR MAXINE, ALEXA AND BEN

PREFACE

During the last 25 years or so, the field of excitable membranes has gone from a position comparable to that of atomic physics at the turn of the century to one of having a substantial body of organized, quantitative information. A number of theories are available on topics related to the overall discipline, and it is possible to work problems in each area. This problem-solving approach is emphasized in this text mainly because of a strong bias on the part of the author that it is only by working with numbers that the student becomes conversant with the dimensions of the field. After working the examples, students will be familiar with the same theoretical models used by professional researchers. It is hoped that this will stimulate some further advances by the readers.

This book was written in order to fill a gap in the available study materials: in timeliness and amount of detail it is somewhere between the available texts and professional review articles. Thus, it would be appropriate for beginning graduate students, medical students, and undergraduates with some experience in calculus, physics, electronics, and chemistry. The necessary elements of any of these may, however, be picked up after starting to study nerve and muscle membranes. The student with a strong interest in the subject will, of course, wish to read the original sources; this volume should provide access to many of the important papers up to the present.

The brief historical review in Chapter 1 may contain some unfamiliar terms, always a problem in studying a new field. It seemed better not to stop and explain each one, as many of the terms would be familiar to readers from other fields and all are covered in detail in other parts of the book. Chapter 2 introduces the major techniques for recording responses of excitable membranes and includes some material on conduction of action potentials in whole nerves. In Chapter 3 the electrical analogue of the membrane is developed and compared with the known behavior of nerve cell bodies, axons, and muscle fibers. Chapter 4 treats the ionic basis of membrane potentials and contains a derivation of the Nernst equation, a method for calculating P_{Na}/P_K ratios, and a sample calculation of alkali-cation selectivities. Chapters 5, 6, and 7 are concerned with the voltage-clamp technique, the Hodgkin-Huxley theory and two other theories of excitation, and the question of independent channels for sodium and potassium currents, respectively. Chapter 8 discusses some of the systems in which action potentials are caused by calcium influx, rather than sodium. The electrogenic properties of metabolic pumps, or the action of these pumps on membrane potential, including a theoretical model, are presented in Chapter 9. Chapter 10 attempts to review several of the innovative approaches to research in the field, some of which may be of major importance before long.

It is hoped that, along with information and skills, the reader may be able to gain some appreciation for the zestful pursuit of research in this field. The excitement of watching the validation of theory by experiment is always amazing. It can even be amazing when the experiments were performed by someone else.

The author was particularly fortunate in being able to prepare this book near the UCLA Biomedical Library. He would also like to thank the colleagues at UCLA who contributed criticism and ideas: Dr. Sergio Ciani, Dr. Sally Krasne, Dr. Susumu Hagiwara, Dr. Gabor Szabo, and John Prehn. The entire manuscript was read and greatly improved by Dr. Charles Leo Ortiz, Dr. Jeffrey Kroin, and Dr. Deric Bownds. Grateful thanks is also extended to Dr. Felix Strumwasser, Dr. Kathy Graubard, Dr. George Moore, Dr. Alvin Siger, Dr. Raphael Gruener, Cathy Hill, Cecilia Ordoñez, Sondra Owens, Dorothy Raffel, and the students of Physiology 224.

—Los Angeles, June, 1975

CONTENTS

NERVE AND MUSCLE EXCITATION

1

DEVELOPMENT OF THE STUDY
OF EXCITATION

The roles which sodium and lithium ions play in the spread of excitation and contraction of muscle are still unclear; perhaps there occurs, during these processes, a certain exchange between the potassium ions of the muscle fiber and the sodium ions in the surrounding solution, but there are important problems with such an assumption.

— OVERTON, 1902

It is difficult to chart the path by which scientific understanding of excitable membranes has progressed, given such prophetic statements as the one above. Overton performed careful experiments on the importance of sodium in the bathing solution for the activation of muscle and prophesied the ionic theory of membrane excitation which is now accepted by almost all electrophysiologists. But the essential correctness of his statement was not proved until almost 50 years later!

THEORY OF TRANSMEMBRANE POTENTIALS

In the field of study of excitable membranes the theory has often outstripped the technology and has had to wait for new technical developments to provide confirmation or denial. By the first decade of the twentieth century, Bernstein (1902, 1912) had produced an extensive theory of membrane permeability changes during excitation in muscle and nerve. Drawing heavily on the electrochemical theories of Nernst (1889), Bernstein tried for the first time to give a physical basis to the activity of living excitable cells. One prediction of his theory was that

1

for a cell membrane separating salt solutions of different concentrations the transmembrane potential should be given by

$$E = \frac{u - v}{u + v} \frac{RT}{F} \ln \frac{[C]_o}{[C]_i} \tag{1}$$

where R = universal gas constant
T = temperature, °K
F = Faraday constant
u = mobility of (+) ions in the membrane
v = mobility of (−) ions in the membrane
$[C]_o$ = external salt concentration
$[C]_i$ = internal salt concentration

Bernstein further proposed that the normal resting potential of excitable cells (inside negative) was due to a selective permeability of the membrane to potassium ions. He believed that the action potential resulted from a general increase in permeability to all ionic species present (mainly Na^+, K^+, and Cl^-), reducing the transmembrane potential to zero. This theory is now testable because measured transmembrane potentials can be compared with Equation 1 with suitable assumptions about the mobilities. But in Bernstein's day two things prevented such a comparison: The internal ion concentrations in nerve and muscle had never been determined, and no one had ever measured a transmembrane potential.

INTRACELLULAR RECORDING OF ACTION POTENTIALS

By 1934, Fenn and others were publishing internal ion concentrations in muscle, but similar data for nerve had to wait for the discovery of the super membrane, the squid giant axon. J. Z. Young started doing experiments related to the structure of these axons about 1936. Then, in 1939 Hodgkin and Huxley published a short note in *Nature* which had an almost breathless air of excitement; in it they described the first direct measurement of a transmembrane potential in a living excitable cell. They observed that during the nerve impulse the transmembrane potential actually *reversed,* and the axoplasm became *positive* with respect to the surrounding solution. This result demanded a modification of Bernstein's theory, which could not account for internal positivity. Aware of the need for data on ionic concentrations in nerve, Bear and Schmitt in 1939 and Steinbach and Spiegelman in 1943 separately published the necessary information for the squid axon. Then the stage was set for the paper in 1949 by Hodgkin and Katz in which they ascribed the resting potential of nerve mainly to a specific potassium permeability, and the action potential to a transient state of the membrane, which was more permeable to sodium than potassium. This paper laid the foundation for the current view of nerve membrane excitation.

ADVENT OF MICROELECTRODES

The generality of the Hodgkin-Katz formulation became evident as intracellular recordings were obtained from other excitable cells. The introduction of electrolyte-filled glass micropipettes by Ling and Gerard in 1949 permitted such recordings in single muscle fibers. This technique was quickly applied to mammalian motoneurones (Brock et al., 1952), toad motoneurons (Araki et al., 1953), invertebrate ganglion cells (Tauc, 1954), and other preparations. Transmembrane potentials were also being recorded in myelinated nerve fibers by insulating between two neighboring nodes of Ranvier with an extracellular partition, and placing one node in isotonic KCl solution, depolarizing it essentially completely. The membrane potential could then be measured between the depolarized and normal nodes (Huxley and Stämpfli, 1951). All these newly accessible membranes appeared to require external sodium for the production of action potentials, and all showed reversal of transmembrane potential during the impulse. The ionic hypothesis of excitation, or the idea that action potentials resulted from the movement of ions across membranes, was coming into its own.

THE MEMBRANE MODEL

By the late 1940s, a considerable amount of effort had been expended on developing an electrical model of an axon which could explain the propagated action potential. It was known that axons did not function simply as longitudinal conductors, or "wires." (The internal resistance was much too high.) Instead, the action potential was propagated *with no loss of amplitude* over great distances through the coupling of activity in one region to a neighboring quiescent region. (This process has been compared to the falling of dominoes arranged in a row, where the "activity" moves continuously from one area to the next.) In order to account for the electrical coupling of adjacent areas of axon membrane, Hodgkin and Rushton (1946) used the CABLE THEORY derived for transatlantic telegraph cables. This method treated the membrane as a series of discrete RC circuits, connected by "internal" resistances (see Figure 6, Chapter 3). This model was very nice in theory, but it contained a serious problem with respect to experimental testing:

Whenever currents flowed *through the membrane,* a significant part of the current was through the capacitive element, i.e., a displacement current not involving the actual movement of ions across the membrane. This difficulty precluded the direct measurement of ionic currents during the development of an action potential.

THE VOLTAGE CLAMP

In 1949, Cole arrived at a solution to the problem, which was elaborated *par excellence* by Hodgkin and Huxley in the early 1950s: In order to eliminate capacitive currents in the squid axon, which were proportional to the rate of change of transmembrane potential, these workers held the transmembrane potential constant during an impulse by means of negative feedback, and measured the amount of injected current necessary to do so. This current was equal and opposite to the active current produced by the membrane. Thus, with potential held constant the capacitive current was eliminated, and pure ionic currents could be measured directly for the first time during the entire nerve impulse.

HODGKIN-HUXLEY MODEL OF SPECIFIC IONIC CHANNELS

An important result of this approach was the discovery of the sequential nature of the transmembrane currents: Hodgkin and Katz had felt that the potassium permeability might not change during the nerve impulse, and indeed no net movement of potassium ions was necessary to explain the shape of the action potential. The voltage-clamp technique revealed that for applied depolarizations above a certain level the initial transmembrane current was inward and that this was followed by a more slowly developing outward current. In a series of experiments, Hodgkin and Huxley showed that the total membrane current was made up of an early inflow of sodium ions, due to a brief increase in sodium permeability, and a later outflow of potassium ions, as a result of a delayed increase in potassium permeability. The mechanisms of sodium and potassium transport were somewhat independent; for instance, removal of external Na^+ did not affect K^+ current. The Hodgkin-Huxley theory treated the two mechanisms as completely independent *sodium* and *potassium channels* in the membrane. More recent experiments — such as the ability of tetrodotoxin to block only inward currents normally carried by sodium, and that of tetraethylammonium to block only outward currents normally carried by potassium — have supported the idea of independent channels (see Chapter 7).

RECENT CHARACTERIZATION OF THE CHANNELS

The studies with the voltage clamp showed the sequential activation of the inward- and outward-current channels, but they gave no suggestion as to *how* ions were actually carried across the membrane during an action potential. This process has been viewed as diffusion of ions through pores in the membrane (Mullins, 1959), as opening of "gates" by rotation of dipoles in the membrane (Goldman, 1964), as an ion-exchange process (Tasaki, 1968), and as storage of sodium in the membrane between impulses (Hoyt and Strieb, 1971).

Most of these are good models, in that they explain the principal relevant facts about membrane currents, potentials, thresholds, time courses, etc. At the same time, their common predictive abilities make it difficult to distinguish one from another, in the attempt to see which best describes a real excitable cell membrane. At least, by thinking in terms of the intimate molecular apparatus involved in excitation we are coming to grips with understanding the process and not just settling for a black-box description of the membrane conductances.

Such a way of thinking also suggests new approaches, such as experiments with fluorescent dyes, with light scattering, and with birefringence, to look for subtle changes in the membrane structure during excitation. Strange pharmacological agents are being found, such as the toxin from a poisonous frog, which selectively increases only the resting sodium permeability of nerve. Workers with artificial membrane systems have been able to induce specific permeabilities by addition of antibiotics and other ion carriers to the membranes. Action potentials have even been produced in controlled artificial systems (see Chapter 10).

The clever combination of these molecular approaches with the known ionic properties of nerve and muscle membranes will one day, probably soon, give us the structures of the natural ion-carrying systems and a detailed understanding of excitation.

REFERENCES

Araki, T., Otani, T., and Furukawa, T. (1953). The electrical activities of single motoneurones in toad's spinal cord, recorded with intracellular electrodes, *Japan. J. Physiol.* **3,** 254–267.

Bear, R. S., and Schmitt, F. O. (1939). Electrolytes in the axoplasm of the giant nerve fibers of the squid, *J. Cell. Comp. Physiol.* **14,** 205–215.

Bernstein, J. (1902). Untersuchungen zur Thermodynamik der bioelektrischen Ströme, *Pflüger's Arch. ges. Physiol.* **92,** 521–562.

Bernstein, J. (1912). *Elektrobiologie.* Braunschweig, Vieweg u. Sohn.

Brock, L. G., Coombs, J. S., and Eccles, J. C. (1952). The recording of potentials from motoneurones with an intracellular electrode, *J. Physiol.* **117**, 431–460.

Cole, K. S. (1949). Dynamic electrical characteristics of the squid axon membrane, *Arch. Sci. Physiol.* **3**, 253–258.

Fenn, W. O., and Cobb, D. M. (1934). The potassium equilibrium in muscle, *J. Gen. Physiol.* **17**, 629–656.

Goldman, D. E. (1964). A molecular structural basis for the excitation properties of axons, *Biophysic. J.* **4**, 167–188.

Hodgkin, A. L., and Huxley, A. F. (1939). Action potentials recorded from inside a nerve fiber, *Nature, Lond.* **144**, 710–711.

Hodgkin, A. L., and Katz, B. (1949). The effect of sodium ions on the electrical activity of the giant axon of the squid, *J. Physiol.* **108**, 37–77.

Hodgkin, A. L., and Rushton, W. A. H. (1946). The electrical constants of a crustacean nerve fibre, *Proc. Roy. Soc. B.* **133**, 444–479.

Hoyt, R., and Strieb, J. D. (1971). A stored charge model for the sodium channel, *Biophysic. J.* **11**, 868–885.

Huxley, A. F., and Stämpfli, R. (1951). Direct determination of membrane resting potential and action potential in single myelinated nerve fibres, *J. Physiol.* **112**, 476–495.

Ling, G., and Gerard, R. W. (1949). The normal membrane potential of frog sartorius fibers, *J. Cell. Comp. Physiol.* **34**, 383–396.

Mullins, L. J. (1959). An analysis of conductance changes in squid axon, *J. Gen. Physiol.* **42**, 1013–1035.

Nernst, W. (1889). Die elektromotorische Wirksamkeit der Ionen, *Zeit. Physik. Chem.* **4**, 129–181.

Overton, E. (1902). Beiträge zur allgemeinen Muskel – und Nervenphysiologie, *Pflüger's Arch. ges. Physiol.* **92**, 346–386.

Steinbach, H. B., and Spiegelman, S. (1943). The sodium and potassium balance in squid nerve axoplasm, *J. Cell. Comp. Physiol.* **22**, 187–196.

Tasaki, I. (1968). *Nerve Excitation: A Macromolecular Approach,* Springfield, Ill., Thomas.

Tauc, L. (1954). Réponse de la cellule nerveuse du ganglion abdominal de *Aplysia* depilans à la stimulation directe intracellulaire, *C. R. Acad. Sci., Paris* **339**, 1537–1539.

Young, J. Z. (1936). The structure of nerve fibres in cephalopods and crustacea, *Proc. Roy. Soc. B.* **121**, 319–337.

2

ELECTRICAL RECORDINGS FROM
NEURONS AND MUSCLE
FIBERS

EXTRACELLULAR TECHNIQUES

Extracellular stimulation and recording techniques were among the
first to be used in electrophysiology, and they are still actively em-
ployed. Therefore, a brief consideration of these methods is relevant to
the study of excitable membranes. A typical arrangement for extracellular studies with an isolated section of nerve is shown in Figure 1. The
nerve is laid over a row of silver wires in a hermetically sealed moist
chamber with a small amount of saline on the bottom. This method
produces much larger signals than recording under the saline solution,
as it avoids shunting of the recorded signals. A pair of wires at one end

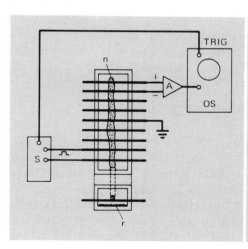

Experimental arrangement
for extracellular recording
of nerve action potentials.
n, nerve; r, Ringer's solution in bottom of chamber;
cross section shown below;
TRIG, trigger signal.

1

2

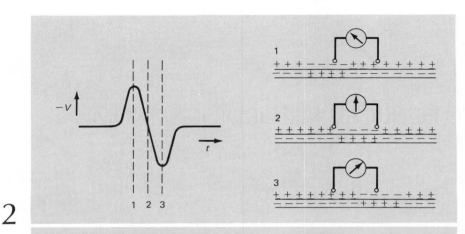

Biphasic action potential. Left: Shape of the extracellular potential when many axons in a nerve are excited synchronously. Right: Explanation of biphasic potential in terms of activity of a single nerve fiber.

of the nerve is connected to a stimulator (S), which produces short voltage pulses to excite the nerve. A pair at the other end is connected to a differential high-gain ac preamplifier (A) which prepares the signal for display on the oscilloscope (OS). The preamplifier must be differential, that is have (+) and (−) inputs, because the nerve signals to be recorded are usually the same size as or smaller than the 60-Hz interference in each lead. By subtracting the signal at the (+) lead from that at the (−) lead one can minimize this interference. The amplifier must have a gain of 100 to 1,000 because the extracellular signals are typically smaller than 1 mV. The amplified signal of 0.1 to 1 V is then easily displayed on most oscilloscopes. Finally, the pre-amplifier must have a high input impedance, or it will shunt the signals and reduce the amplitude. Typical input impedances for biological preamplifiers of this sort are greater than 100 MΩ.

THE COMPOUND ACTION POTENTIAL

The sequence of events during stimulation of the nerve and display of the extracellular action potential is: (1) The stimulator emits a brief trigger pulse which starts the oscilloscope beam moving horizontally across the screen. (2) After a certain delay, the stimulator emits a square stimulus pulse (may be 0.1 to 2.0 msec, 1 to 10 V). As the amplitude of the pulse is increased, it excites some of the axons which make up the piece of nerve. (3) The action potential produced is picked up

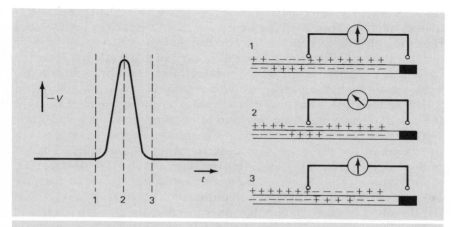

3

Monophasic action potential obtained with killed-end recording. Left: Extracellular potential recorded from many axons. Right: Explanation based on activity of a single fiber.

by the recording electrode pair and produces a trace on the oscilloscope screen such as that shown on the left of Figure 2. This is called a COM-POUND ACTION POTENTIAL because it is made up of the summed activity of hundreds of axons. We can understand the shape of this action potential by considering the activity of a single axon, as shown in Figure 2, right. The resting distribution of charges across the axon membrane is negative inside, as shown on the left side of the drawings. At time 1 the nerve impulse, consisting of a region of reversed polarity (positive inside), reaches the first external recording electrode. This causes the negative deflection seen at the top of the figure. At time 2 the impulse has moved along the axon to a point halfway between the recording electrodes; so the recorded potential returns to zero. At time 3 the impulse has reached the second electrode, and a positive deflection is seen. This entire event is usually referred to as BIPHASIC ACTION POTENTIAL. The compound action potential recorded outside the whole nerve (left) results from hundreds of such miniature spikes occurring almost synchronously. The individual axon spikes are ALL-OR-NONE, that is, they will occur in exactly the same stereotyped way each time the stimulus exceeds the THRESHOLD level (which varies from axon to axon). The compound action potential, on the other hand, is GRADED and varies in size with the stimulus amplitude. This is because increasing the stimulus brings more and more axons into play, each of which contributes to the total record.

In another method of extracellular recording, one recording electrode is placed on a crushed end of the nerve, which is therefore unable to conduct. This produces a constant INJURY CURRENT between the recording electrodes because the crushed end is negative with respect to the uninjured portion. However, this current is usually not recorded by the ac preamplifiers used for extracellular recording. As the impulse passes down the nerve, an action potential such as that in Figure 3, left is recorded. The entire event has a negative polarity because the positive phase has been removed by crushing the nerve under one electrode. This can be understood if we consider the behavior of one of the axons in the nerve, as shown on the right of the figure: The recorded signal is always negative as the impulse (1) approaches, (2) encounters, and (3) passes the active recording electrode. Because the crushed end is inactive, the impulse cannot pass by it and reverse the direction of current flow between the electrodes as in Figure 2, right. This technique is often called KILLED-END RECORDING, and the action potentials recorded by this method are called MONOPHASIC.

Nerve membranes also have the property of REFRACTORINESS following an action potential: For a short time (less than 1 msec) following a spike in a single axon, no second action potential can be elicited, even if the stimulus is greatly increased. This is known as the ABSOLUTE REFRACTORY PERIOD. A slightly longer time after the first action potential, a second one can be produced if the stimulus is made larger than that needed to elicit the first spike. This is the RELATIVE REFRACTORY PERIOD. In whole nerves, the refractoriness of single axons is seen in the following way: As two stimuli occur closer and closer together in time, more and more single axons become refractory, and the second compound action potential is *gradually* reduced to zero.

THE FROG NERVE

So far we have considered the activity of only a single group of axons in the nerve, all of which have about the same conduction velocity; thus the individual axonal action potentials were conducted from the stimulating site to the recording site at about the same rate and could sum to give distinct compound action potentials. However, almost all peripheral nerves have more than one group of fibers with particular conduction velocities. For example, in the frog peroneal nerve at least three different peaks can be resolved in the compound action potential (Figure 4) obtained with killed-end recording. As the stimulus amplitude is increased, the first peak to appear is that labeled α. As the stimulus amplitude is further increased first the β and then the γ deflections appear. This is because the fibers giving rise to the α peak have the lowest threshold. They also have the largest conduction velocity, as the

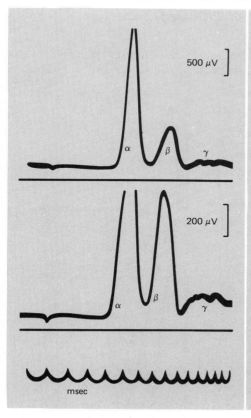

500 μV

200 μV

msec

Compound action potential from frog peroneal nerve, obtained with killed-end recording. α, β, and γ peaks due to different conduction velocities of groups of axons in the nerve. (Time scale compressd on right.) (Erlanger and Gasser, 1937.)

4

α peak appears earliest after the stimulus artifact (small downward deflection). This illustrates the general rule-of-thumb that *fibers with the lowest threshold have the greatest conduction velocity.*

In Figure 5 is shown Erlanger and Gasser's demonstration that the α and β peaks represent the activity of groups of fibers having different conduction velocities. Killed-end recording was used at one end of the nerve, and closely spaced pairs of stimulating electrodes were placed at various distances from the crushed end, indicated on the left margin. When the stimulating pair was very close to the recording electrode (top trace), the α and β deflections overlapped. As the conduction distance was increased, the two deflections became more and more separated in time. This could only have been because the fiber groups giving rise to each deflection had different (but constant) conduction velocities. The slanting lines are drawn through the estimated beginning of the α and β deflection in each trace. The steeper α line indicates the greater conduction velocity of the α group of fibers.

VARIATION OF CONDUCTION PARAMETERS WITH FIBER DIAMETER

From measurements on histological sections of nerves, it is possible to construct histograms of the diameters of the fibers present. Such a

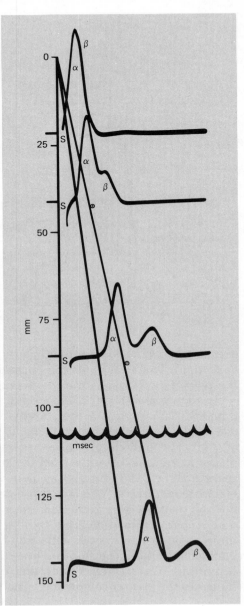

Demonstration that α and β peaks represent the activity of fiber groups with different conduction velocities. Numbers to left are distances in millimeters from stimulating electrode pair to recording electrode. (s indicates time of stimulus.) (Erlanger and Gasser, 1937.)

5

histogram for a human sensory nerve is shown in the inset of Figure 6. Two peaks are clearly present in the distribution of diameters; one has a mean of about 3 μm, and the other has a mean of about 12 μm. The compound action potential for this nerve has the form shown in the graph of Figure 6, with two well-defined elevations. Erlanger and Gasser showed in several studies that the *first*, or *fastest-conducted*, elevation, was due to the *large-diameter* fiber group. Erlanger and Gasser made hundreds of reconstructions of action potentials by making various assumptions about the contributions of individual fiber groups to the total record. The first, and most necessary assumption, was that *the conduction velocity of each fiber was proportional to its outside diameter*. By delaying the contributions of each size of fibers by the appropriate conduction time and summing, they could obtain a fair representation of the compound action potential. Erlanger and Gasser were able to obtain much better fits to the compound action potential assuming that *individual spike amplitudes were proportional to fiber diameters*. These assumptions were also found to hold for cat neurons (Hursh, 1939).

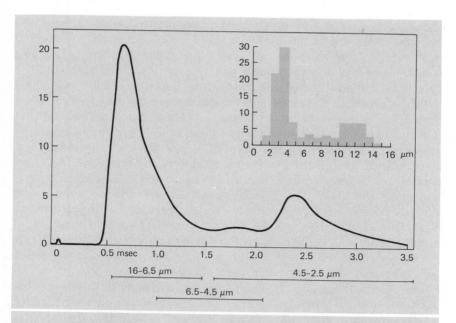

6

Compound action potential and fiber-diameter histogram for human sensory nerve. Ordinate scale in arbitrary units. Stimulus applied at time 0. Ranges of fiber diameters giving rise to each part of action potential indicated below. (Gasser, 1943.)

7

Myelinated nerve fiber, showing outward current at one node produced by activity in a neighboring node.

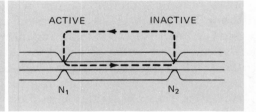

THE NODE OF RANVIER

The explanation of monophasic and biphasic action potentials in Figures 2 and 3 applies, strictly speaking, only to simple cylindrical axons. Real nerves almost always contain a large proportion of MYELINATED fibers. These are fibers with a multilayered coat of myelin membranes, longitudinally interrupted every 1 to 2 mm by a constriction called the NODE OF RANVIER. Here the insulating myelin disappears, and the central axon is exposed to the external solution, as diagrammed in Figure 7.

Tasaki and Takeuchi (1941 and 1942) first showed that propagation of the nerve impulse in myelinated fibers occurs by SALTATORY CONDUCTION (from the Latin *saltare,* to dance). This process is illustrated in Figure 8, from the work of Huxley and Stämpfli (1949). The frog sciatic nerve was dissected down to a single myelinated fiber, and the fiber was pulled through an insulating barrier so that potentials could be recorded from one side of the barrier to the other. The recorded potentials were proportional to the average longitudinal current inside the fiber between the two sides of the barrier and inversely proportional to the longitudinal resistance of the fiber. By taking the difference of longitudinal currents at different regions of the fiber, Huxley and Stämpfli found the *membrane current* as a function of position along the fiber. (The Y shapes on the right indicate the position of the barrier at which the records were calculated.) Because the nerve was stimulated at the same location each time, the action currents occurred earlier and earlier as the barrier was moved along the fiber. It can be seen that only very small currents were recorded in the internodal regions, while large currents were recorded near the nodes. The action potential apparently "hopped" from one node to the next. This is known to result from the excitation of an inactive node by activity in a neighboring node, as shown in Figure 7. The action potential at N_1 results from an inward flow of sodium ions (to be discussed in Chapters 4 and 5). This results in an outward current at N_2, which excites the inactive node and produces an action potential there. About 0.1 msec after an action potential at one node, a new action potential suddenly appears at the next node, with no activity in the internodal region. This method of conduction is to be contrasted with that in un-myelinated axons, where the action potential moves smoothly, without jumps. (However, conduction in *demyelinated* fibers, where the myelin sheath has been removed by application of diphtheria toxin, is saltatory; Rasminsky and Sears, 1972.) The mechanism of propagation in unmyelinated axons will be discussed further in Chapter 3.

In 1951, Huxley and Stämpfli came up with an ingenious method to measure the transmembrane potential of a single node, which is illustrated in Figure 9. The nodes N_1 and N_2 were isolated by two high-resistance gaps (g) in the external medium (one was filled with paraffin oil and the other was a thin layer of celluloid). Node N_1 was placed in normal Ringer's solution (an artificial solution similar to the ex-tracellular body fluid) and node N_2 was placed in isotonic KCl, which depolarized it completely. One might think that the membrane resting potential could now be measured directly between N_1 and N_2, because

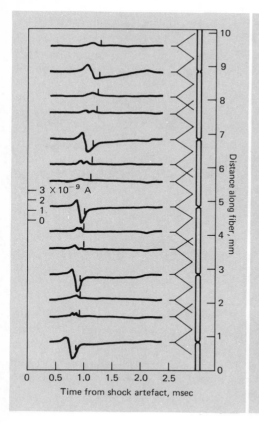

Saltatory conduction in myelinated fiber. Traces show currents recorded at various positions along the fiber. (Huxley and Stämpfli, 1949.)

8

N_1 had a normal resting potential E_m, and N_2 had none. However, even the small amount of saline solution under the gaps was enough to shunt, and greatly reduce, the measurable potential. Huxley and Stämpfli avoided this problem by measuring the short-circuit current flowing through the axoplasmic resistance (R) and under the insulating barriers in the path N_1N_2CBA. They measured this current as a voltage drop across the high barrier resistance from B to C. Then an external voltage V was applied between A and B and adjusted to reduce the current to zero. This value of V was presumably equal and opposite to E_m, the potential giving rise the short-circuit current. The average resting potential found by these investigators was -71 mV.

The method was even adapted to measurement of action-potential amplitudes; pulses of voltage which just reduced the short-circuit current to zero were applied from A to B at the peak of the action potential. The average value of overshoot measured in this way was $+45$ mV.

THE SUCROSE GAP

The successful recording of membrane potentials at the node of Ranvier led to the clever use of an extracellular insulating barrier for recording membrane potentials in whole axons or bundles of axons. In

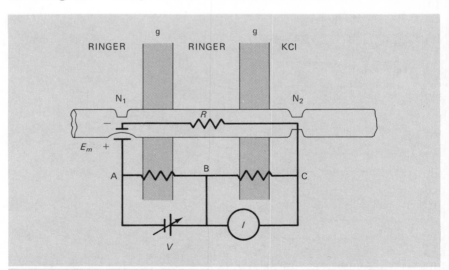

9

Potentiometric method of measuring transmembrane potentials in single node of Ranvier. V is adjusted until $I = 0$; then $V = -E_m$. (After Huxley and Stämpfli, 1951.)

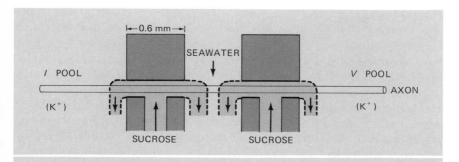

The sucrose-gap method of measuring membrane potentials in axons. (Julian *et al.*, 1962*a.*)

1954, Stämpfli first described the method, and it was developed considerably by Julian *et al.* (1962*a, b*) using lobster axons. As shown in Figure 10, the insulating barriers are streams of sucrose flowing around the axon in such a way as to create three isolated sections of nerve. The two end sections are depolarized with KCl, and the central section is perfused with normal saline, thus becoming an "artificial node." Membrane potentials can be measured from one KCl pool to the saline pool, and currents may be injected via the other KCl pool. Stämpfli (1954) found that flowing sucrose was necessary to the success of this method, because petroleum jelly or another stationary insulator did not remove the ions adherent to the axons under the insulator. The sucrose-gap method has been applied to many studies of nerve properties, especially with axons which are too small to impale with longitudinal electrodes, as described in the next section.

INTRACELLULAR RECORDING

The development of the field was arrested for more than 4 years during World War II. In 1945, Hodgkin and Huxley finally were able to continue their studies, begun in 1939, with direct intracellular measurements of resting and active membrane potentials in the squid giant axon. To make these observations, they had to use a high-impedance amplifier (10^{10} Ω) so as to not shunt the recorded signals. However, unlike with extracellular techniques, the gain could be low, because the intracellular potentials were quite large (50 to 100 mV). Also, the amplifier was a dc (direct-coupled) type, instead of the ac type used for extracellular measurements, so that they were able to amplify constant potential differences such as the resting potential. (These techniques were also used for dc measurements with the sucrose gap.)

The early capillary electrodes which Hodgkin and Huxley used to record transmembrane potentials were about 100 μm in diameter and were introduced vertically into the cut end of a suspended length of axon. The glass capillary electrode was filled with seawater, and a silver wire ran inside along most of the length. One lead of the preamplifier was connected to the silver wire, and the other amplifier lead was connected to a coil of chlorided silver in the external solution. Stimuli were applied by two platinum wires applied to the outside of the axon. A typical intracellular action potential recorded by this method is shown in Figure 11. The resting potential is about −45 mV, and the OVERSHOOT, or size of the positive phase measured from zero, is about +40 mV. The time base indicates that the action-potential duration is about 0.7 msec. Also evident is the AFTER POTENTIAL, which follows the positive-going part of the spike; in the after potential the membrane potential is HYPERPOLARIZED, or made more negative than the resting potential, for about 2 msec. Action potentials in these single axons are ALL-OR-NONE; that is, once the stimulus reaches a certain level the action potential will occur, and no further increase in stimulus amplitude will change the form of the response.

In 1949, a more direct method of measuring membrane potentials in single muscle fibers was developed: Ling and Gerard pulled fine glass micropipettes, filled them with isotonic KCl, connected them to the recording amplifier, and stuck the electrodes right into living fibers. If the electrode tips were small enough (<0.5 μm), then membrane potentials could be measured for up to several hours. This

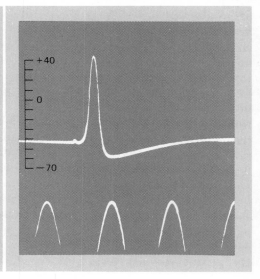

11 Intracellular recording of action potential from squid axon. Time base: 500 Hz. (Hodgkin and Huxley, 1945.)

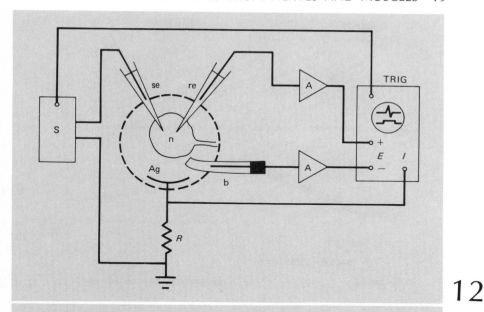

12

Experimental arrangement for intracellular recording from nerve or muscle fiber. n, nerve or muscle cell; E, transmembrane potential; TRIG, trigger signal. See text for more details.

method was soon developed by Nastuk and Hodgkin (1950) for muscle fibers, and applied to many other preparations. A typical experimental arrangement for microelectrode work is shown in Figure 12. The preparation (in this case, a nerve cell body) is placed in a small recording chamber and covered with normal saline solution (a solution which approximates the natural extracellular fluid of the animal). Stimulating and recording electrodes (se and re) having resistances of 5 to 20 MΩ are inserted into the cytoplasm through the cell membrane. A bridge (b) filled with saline-agar gel serves as the external voltage reference electrode. In the gel lies the chlorided silver lead, which is thus protected from bumps and scrapes and changing ion concentrations around the preparation. Membrane potentials are measured differentially on the oscilloscope by subtracting the outputs of the pre-amplifiers (A). These preamplifiers must be direct-coupled and should have input impedances of at least $10^{10}\Omega$. The stimulator (S) applies current (I) between the stimulating electrode and ground. The return path for the current is through a chlorided silver wire in the bath (Ag) and series resistor (R). Injected current may thus be measured as the voltage drop across R.

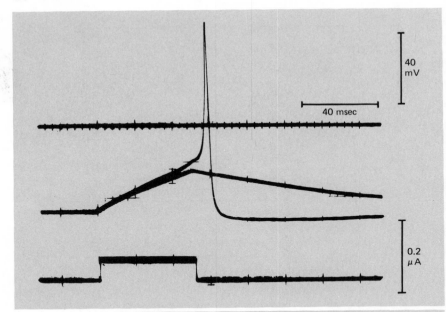

13

Action potential in an *Aplysia* neuron cell body. Upper trace, zero potential; middle trace, membrane potential; lower trace, current injected through membrane with intracellular microelectrode. Two sweeps superimposed, one in which the stimulus is subthreshold and one in which it is suprathreshold.

To observe membrane responses with this arrangement, a trigger pulse is first produced, which starts the oscilloscope beam moving. Then a square stimulus pulse is applied to the cell. This produces a response such as that shown in Figure 13, taken from an *Aplysia* ganglion cell. The top trace shows the zero of potential, before the recording electrode was placed in the cell; each middle trace is membrane potential, and the bottom trace is injected current. Superimposed sweeps are shown, one in which the stimulus was too small to excite, and one in which it was just large enough. The *threshold* depolarization can be measured as that which is just large enough to excite. The resting potential was −48 mV, and the overshoot was +56 mV. (How simple measurements such as these now seem, with our commercially available amplifiers and stimulators. Such data did not come as easily to the workers in the 1940s, but it was often first-rate in quality; c.f. Figure 11.)

This has been a quick overview of extracellular and intracellular recording techniques. Both methods have been extended to many new situations. For instance, extracellular recordings of single-neuron activity have been obtained almost everywhere in the brain of the cat. Intracellular recordings have been made in single spinal motoneuron cell bodies (Brock *et al.*, 1952) and even in the cerebral cortex (Albe-Fessard and Buser, 1953). In each case, however, the basic strategies of high-gain ac amplifiers for extracellular work and low-gain dc amplifiers for intracellular work have been followed.

PROBLEMS

1. In Figure 5, calculate the conduction velocities of the fiber groups giving rise to the α and the β peaks in the compound action potential.

2. In Figure 6, the conduction distance was 4 cm. Calculate the conduction velocities for the two peaks from the time between the stimulus artifact (small upward deflection at time 0) and the *start* of each elevation.

3. Using the conduction velocities in Problem 2 and the mean fiber diameters of each peak in the histogram in Figure 6, plot a two-point curve of velocity versus diameter. Estimate the slope of a straight line relating velocity and diameter.

4. The duration of the α deflection of the monophasic action potential in Figure 4 (about 1 msec) represents the time taken by the *wave* of activity in the nerve to pass by the recording electrode. If the wave is moving with the velocity found in Problem 1, what *length* of nerve is active during the action potential (or what is the wavelength of the action potential)?

5. In Figure 12, how large should the current-measuring resistor R be to give a sensitivity on the current channel of the oscilloscope of a 10-mV deflection for a 100-nA injected current?

REFERENCES

Albe-Fessard, D., and Buser, P. (1953). Explorations de certaines activités du cortex moteur du chat par microeléctrodes: dérivations endosomatiques, *J. Physiol. Path. Gen.* **45**, 14–16.

Brock, L. G., Coombs, J. S., and Eccles, J. C. (1952). The recording of potentials from motoneurones with an intracellular electrode, *J. Physiol.* **117**, 431–460.

Erlanger, J., and Gasser, H. S. (1937). *Electrical Signs of Nervous Activity,* Philadelphia, University of Pennsylvania.

Gasser, H. S. (1943). Pain-producing impulses in peripheral nerves, *Assoc. Res. Nerv. Ment. Dis., Proc.* **23**, 44–62.

Hodgkin, A. L., and Huxley, A. F. (1945). Resting and action potentials in single nerve fibres, *J. Physiol.* **104**, 176–195.

Hursh, J. B. (1939). Conduction velocity and diameter of nerve fibers, *Amer. J. Physiol.* **127**, 131–139.

Huxley, A. F., and Stämpfli, R. (1949). Evidence for saltatory conduction in peripheral myelinated nerve-fibres, *J. Physiol.* **108**, 315–339.

Huxley, A. F., and Stämpfli, R. (1951). Direct determination of membrane resting potential and action potential in single myelinated nerve fibres, *J. Physiol.* **112**, 476–495.

Julian, F. J., Moore, J. W., and Goldman, D. E. (1962a). Membrane potentials of the lobster giant axon obtained by use of the sucrose-gap technique, *J. Gen. Physiol.* **45**, 1195–1216.

Julian, F. J., Moore, J. W., and Goldman, D. E. (1962b). Current-voltage relations in the lobster giant axon membrane under voltage clamp conditions, *J. Gen. Physiol.* **45**, 1217–1238.

Ling, G., and Gerard, R. W. (1949). The normal membrane potential of frog sartorius fibers, *J. Cell. Comp. Physiol.* **34**, 383–396.

Nastuk, W. L., and Hodgkin, A. L. (1950). The electrical activity of single muscle fibers, *J. Cell. Comp. Physiol.* **35**, 39–73.

Rasminsky, M., and Sears, T. A. (1972). Internodal conduction in undissected demyelinated nerve fibers, *J. Physiol.* **227**, 323–350.

Stämpfli, R. (1954). A new method for measuring membrane potentials with external electrodes, *Experientia* **10**, 508–509.

Tasaki, I., and Takeuchi, T. (1941). Der am Ranvierschen Knoten entstehende Aktionsstrom und seine Bedeutung für die Erregungsleitung, *Pflüger's Arch. ges. Physiol.* **244**, 696–711.

Tasaki, I., and Takeuchi, T. (1942). Weitere Studien über den Aktionsstrom der markhaltigen Nervenfaser und über die elektrosaltatorische Übertragung des Nervenimpulses, *Pflüger's Arch. ges. Physiol.* **245**, 764–782.

3

THE MEMBRANE ANALOGUE

ELECTRICAL PROPERTIES OF NEURON CELL BODIES

To describe the behavior of biological membranes, it is often convenient to employ electrical models, or analogues. These are approximations to the real membranes, but they have well-defined properties. Thus, we can say a resting nerve cell body behaves like a single-section RC circuit in response to an applied current. This is illustrated in Figure 1. The top shows the voltage response of an *Aplysia* nerve cell body to a subthreshold current pulse lasting a little over 1 sec. The resting potential is about -45 mV. Although the pulse of current is square, that is, the current is applied almost instantaneously, the curve of potential takes about 1 sec to reach its final value, near -39 mV. After the end of the pulse, a similar period of time is required for the potential to return to the resting level.

An electrical analog which has the same type of response is shown in Figure 2. E_m is the total membrane potential at any time; E_r is the RESTING POTENTIAL; R_m is the membrane RESISTANCE; and C_m is a parallel membrane CAPACITANCE. If a current $I(t)$ is applied to this circuit, it will divide into two components: A current I_r will flow in the resistive branch, and I_c through the capacitative element, such that

$$I(t) = I_r + I_c \tag{1}$$

From Ohm's law,

$$I_r = \frac{1}{R_m}(E_m - E_r) \tag{2}$$

23

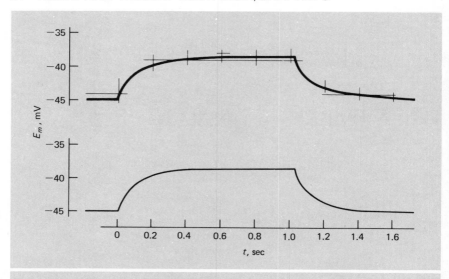

1

Comparison of nerve membrane response to square current pulse and that of a single-section RC filter. Top: Membrane potential of *Aplysia* neuron cell body. Bottom: Exponential response of a single-section filter designed to match the top response. Current step applied at time 0.

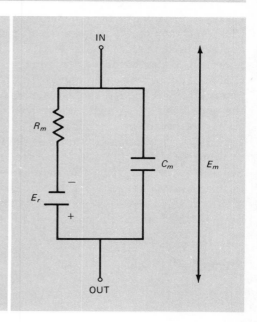

2

Single-section model of resting neuron cell body. The meaning of the abbreviations is found in the text.

The charge on the capacitor is given by

$$q = C_m E_m \tag{3}$$

Differentiating both sides of Equation 3 gives

$$\frac{dq}{dt} = \frac{d}{dt}(C_m E_m) \tag{4}$$

But dq/dt is the rate of flow of charge in the capacitative element, or I_c. So

$$I_c = C_m \frac{dE_m}{dt} \tag{5}$$

From Equations 1 and 2,

$$I(t) = \frac{1}{R_m}(E_m - E_r) + C_m \frac{dE_m}{dt} \tag{6}$$

The current stimulus is zero until time $t = 0$; then it has an amplitude i until the end of the pulse. These end conditions mean that the solution to Equation 6 for the rising phase of the voltage response is

$$E_m - E_r = iR_m(1 - e^{-t/\tau}) \qquad \tau = R_m C_m \tag{7}$$

$E_m - E_r$ is the displacement of potential from resting, in this case a depolarization. The final, or asymptotic value of this depolarization is iR_m. The membrane TIME CONSTANT, τ, determines how rapidly E_m rises after the start of the stimulus. For instance, when $t = \tau$,

$$E_m - E_r = iR_m(1 - e^{-1}) = 0.632iR_m \tag{8}$$

By this time, the depolarization has reached 63.2% of the final value. The time constant for the response shown in Figure 1 is 0.124 sec. After several time constants have passed, the depolarization is quite close to iR_m.

After the end of the current pulse, the equation for the falling phase of the voltage response is

$$E_m - E_r = iR_m e^{-t/\tau} \tag{9}$$

Equations 7 and 9 are plotted at the bottom of Figure 1 for comparison with the real membrane response.

Although the definitions R_m, C_m, E_m and τ apply, strictly speaking, only to the membrane analogue, they are freely transferred to the real cell for purposes of description. The usual method of measuring membrane resistance in a cell body is to apply a known current, i, for a long enough time that the potential change is essentially complete. Then

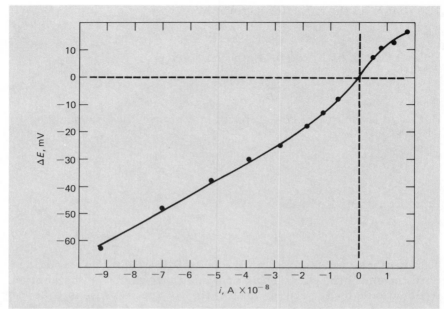

Resting current-voltage relationship of *Aplysia* neuron. Current steps of amplitude *i* applied through second intracellular electrode. ΔE measured from resting potential.

R_m may be calculated as $(E_m - E_r)/i$. The time constant, τ, may be measured as the length of time required for $E_m - E_r$ to reach 63.2% of iR_m. The capacity, C_m, can then be calculated as τ/R_m.

CURRENT-VOLTAGE RELATION

This model is really only a first approximation to the resting cell membrane because R_m is not constant in most excitable cells but rather varies with membrane potential. This is a general electrical property called RECTIFICATION; it is illustrated in Figure 3. The abscissa shows the amplitude of current steps applied to an *Aplysia* cell body, and the ordinate is the resulting change in potential. As the membrane is hyperpolarized (made more negative) or depolarized (made more positive) further than about 10 mV from resting, the curve starts to bend. If we remember that

$$R_m = \frac{E_m - E_r}{i} = \frac{\Delta E}{i} \tag{10}$$

then it is clear that R_m *decreases* from its value near the resting potential as the cell is strongly depolarized or hyperpolarized, because $\Delta E/i$ becomes smaller at potentials further than about 10 mV from resting. Near the resting potential, however, it is a good approximation to say that R_m is constant.

Almost all excitable cells show some sort of rectification over certain regions of membrane potential. For example, the squid axon shows a constant resistance in the hyperpolarizing direction and a relatively constant, but much lower, resistance in the depolarizing direction (Hodgkin, Huxley, and Katz, 1952). This behavior is known to result from an increase in potassium conductance with maintained depolarization. In frog skeletal muscle, however, the membrane resistance *increases* with depolarization (Adrian and Freygang, 1962). This property is called ANOMALOUS RECTIFICATION and is also seen in certain cardiac muscle fibers (Noble, 1965) and in squid axons injected with tetraethylammonium ions (Armstrong and Binstock, 1965). In these cases, the phenomenon has been related to a *decrease* of potassium conductance with depolarization. The *I-V* curve for the *Aplysia* cell body is not exactly like that of the squid axon or frog muscle but is S-shaped. This type of behavior is also sometimes referred to as anomalous rectification (Kandel and Tauc, 1966).

CABLE PROPERTIES OF AXONS AND MUSCLE FIBERS

The model in Figure 2 gives a good approximation of the behavior of cell bodies, because these are practically isopotential inside the membrane; that is, a microelectrode placed anywhere inside a neuron soma

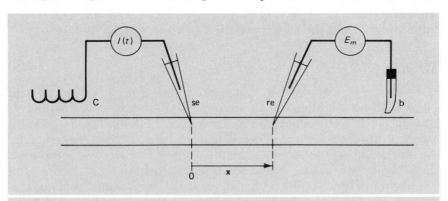

4

Experimental arrangement for stimulating and recording in different regions of an axon or muscle fiber. Text contains the symbol definitions.

records essentially the same voltage with respect to the outside. However, this model cannot describe the behavior of an axon or vertebrate muscle fiber, where the transmembrane potential may vary from point to point. This situation is diagrammed in Figure 4: The stimulating electrode (se) is impaled into the axon or muscle fiber at the point $x = 0$, and a current $I(t)$ is passed between this electrode and an external silver coil return electrode (c). The recording electrode (re) is impaled at varying distances from se. The membrane potential, E_m, is measured between re and an external agar bridge (b).

If a current step is applied at $x = 0$, then the resulting depolarization has a peak at $x = 0$, and falls off with distance from se. This is shown in Figure 5, part A, at various times after the start of the stimulus (Hodgkin and Rushton, 1946). V_m is equal to E_m minus the resting potential. The values of X shown are equal to x divided by the length constant λ (to be defined). Part B shows the decrease in potential with distance from se after the end of the stimulus. In part C is shown the time course of buildup of depolarization at various distances from se. Clearly, the responses develop much more slowly away from se than near it. Part D shows that time course of potential after the stimulus at various distances from se. Values of T are equal to t divided by the time constant τ. Parts C and D are just another way of plotting the same data as shown in parts A and B.

These curves illustrate the process of ELECTROTONIC CONDUCTION. In an axon or muscle fiber, potential changes in one region are conducted to other parts of the cell by a purely passive mechanism. This process is DECREMENTAL; i.e., the signals fall off with distance, unlike the propagated action potential, which has a relatively constant amplitude. Electrotonic signals are also *graded* and vary with the stimulus amplitude, while the action potential is all-or-none.

CABLE THEORY

The simplest theory which can explain this behavior of axons and muscle fibers is called CABLE THEORY, as it was first worked out for transatlantic telegraph cables by Lord Kelvin (1855). In this view, the axon or muscle fiber is considered as a conductive cylinder surrounded by an insulating dielectric layer, the whole structure being immersed in a highly conductive medium. The electrical model of such a cable is shown in Figure 6. In the model, identical "membrane" sections consisting of a resting potential E_r, membrane resistance r_m, and membrane capacitance c_m, are connected together by "internal" resistors r_i. The individual sections are considered to be very small and close together, so that the parameters r_m, r_i, and c_m are distributed evenly

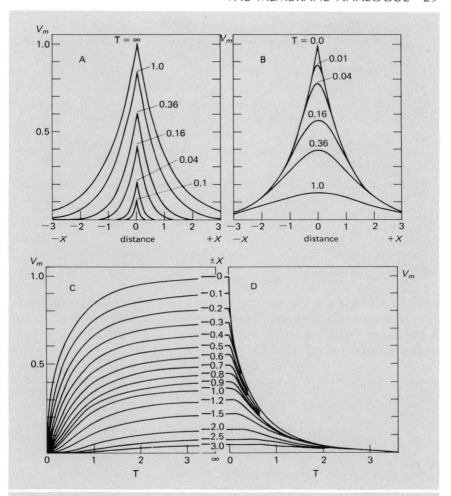

Calculated electrotonic responses of a nerve fiber. Step current applied at
$X = 0$. Part A: Falloff of resulting depolarization with distance from point
of stimulation at various times after onset of stimulus. Part B: Spatial
distribution of potential after end of stimulus. Part C: Time course of
depolarization produced at various locations. Part D: Time course of potential changes after end of stimulus. V_m given as a fraction of the maximum value. (Hodgkin and Rushton, 1946.)

5

along the cable. The resting potential E_r is in millivolts. The other
parameters apply to a 1-cm length of cable; r_m is in ohm-centimeters,
r_i is in ohms per centimeter, and c_m is in farads per centimeter.

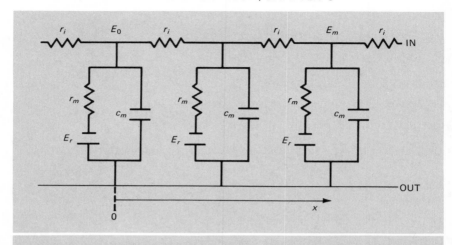

6

Electrical model of a cable applicable to axons or muscle fibers. For details, see the text.

In Figure 6, E_0 is the potential difference from inside to outside at the point $x = 0$, somewhere along an infinite cable. E_m is the potential at any point x and time t. The differential equation governing E_m may be derived as (Hodgkin and Rushton, 1946)

$$-\lambda^2 \frac{\delta^2 E_m}{\delta x^2} + \tau_m \frac{\delta E_m}{\delta t} + E_m - E_r = 0 \tag{11}$$

where

$$\lambda = \sqrt{r_m/r_i} \text{ is the membrane } \textit{length constant}$$

$$\tau_m = r_m c_m \text{ is the membrane } \textit{time constant}$$

If a maintained current is applied to the core of the cable (as with a microelectrode) at the point $x = 0$ and starting at $t = 0$, then the voltage response of the cable at $x = 0$ is (Hodgkin and Rushton, 1946; Rall, 1960)

$$E_m - E_r = \Delta E_0 \operatorname{erf} \sqrt{t/\tau_m} \tag{12}$$

where E_0 = steady-state depolarization at $x = 0$
       erf is a tabulated function
This response is illustrated in Figure 5, part C, for $x = 0$.

When a sufficiently long time has passed after application of the current at $x = 0$, then a steady-state distribution of potential is reached along the cable. At this time, $\delta E_m/\delta t = 0$; so Equation 11 becomes

$$-\lambda^2 \frac{d^2 E_m}{dx^2} + E_m - E_r = 0 \tag{13}$$

The solution to this equation is

$$E_m - E_r = \Delta E_0 e^{-x/\lambda} \tag{14}$$

where ΔE_0 = steady-state depolarization at $x = 0$

In other words, the steady-state distribution of potential along a cable falls off exponentially with distance from the point of injection of current. This is shown in Figure 5, part A, for $t = \infty$.

PROPAGATION OF THE ACTION POTENTIAL

It is this electrotonic coupling of nearby regions of an axon or muscle fiber which is responsible for propagation of the action potential: the active (depolarized) region for a brief period may have a transmembrane potential which is 100 mV more positive than resting. This depolarization is conducted to the adjacent inactive area of the membrane by the cable-type conduction described. The sufficiency of this method of propagating the action potential was shown by Hodgkin in 1937; he demonstrated that propagated action potentials which arrived at a crushed portion of nerve could depolarize the inactive nerve beyond the blocked area and could excite the inactive region if the block were small enough.

SOMA-AXON COUPLING IN REAL NEURONS

In real nerve cells, the electrical analogue which best describes the observed responses of the resting membrane is neither that of a cell body nor of an axon; it is both. In 1960, Rall described the theory for a lumped RC cell-body model connected to one end of an infinitely long (axon) cable. When a constant current is injected into the cell body with attached axon, the voltage response in the cell body is

$$E_m - E_r = \frac{\Delta E_0}{\alpha - 1}\left[\alpha\,\text{erf}\sqrt{\frac{t}{\tau}} - 1 + e^{(\alpha^2 - 1)t/\tau}\,\text{erfc}\,\alpha\sqrt{\frac{t}{\tau}}\right] \tag{15}$$

where ΔE_0 = steady-state potential change in cell body
α = ratio of axon end conductance to soma conductance
τ = time constant
erf and erfc are tabulated functions

This formidable expression may be simplified and applied to real nerve cells if we consider the significance of the parameter α: This number indicates the degree of dominance of the axon over the soma in giving rise to a certain voltage response. When $\alpha = 0$, the condition of very low axon end conductance or no coupling of the soma to the axon, then

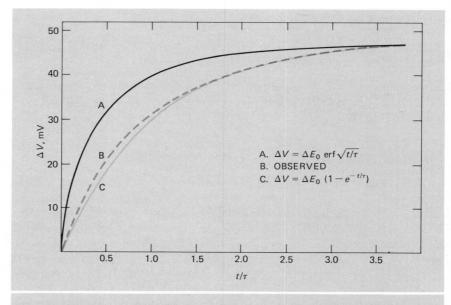

A. $\Delta V = \Delta E_0 \ \text{erf} \sqrt{t/\tau}$
B. OBSERVED
C. $\Delta V = \Delta E_0 \ (1 - e^{-t/\tau})$

7

Analysis of degree of cable-like properties of *Aplysia* cell based on Rall theory. Curve A: Potential response of a semi-infinite cable to a current step. Curve C: Response of a single-section RC filter to the same stimulus. Curve B: Response of *Aplysia* cell. ($\Delta E_0 = 47.3$ mV.)

Equation 15 reduces to

$$E_m - E_r = \Delta E_0(1 - e^{-t/\tau}) \tag{16}$$

which is just the response of a single-section filter (Equation 7). When $\alpha \to \infty$, the condition of complete dominance by the axon, then Equation 15 becomes

$$E_m - E_r = \Delta E_0 \ \text{erf} \sqrt{\frac{t}{\tau}} \tag{17}$$

as in Equation 12. Thus, by examining the shape of the charging curve of a neuron cell body it is possible to estimate the degree of coupling of the cell body to axon or dendrite cable-like structures.

In Figure 7 the voltage response of an *Aplysia* neuron cell body (B) to an injected constant current is compared with Equation 16 (C) and Equation 17 (A). It is evident that α is quite small and that the cell body itself dominates the response. This situation should be contrasted with motoneurons in the cat spinal cord, which have a large amount of attached dendrites with cable-like properties. Here, the value of α

may be 25 or more, and the responses recorded in a cell body closely resemble those of a cable (Rall, 1959).

These electrical analogues of membranes have served a useful function in formalizing our ideas about nerve and muscle cells. However, it should be emphasized that they are based on constant parameters (i.e., time- and voltage-invariant) and are thus an unrealistic view of living systems. In the following chapters, we shall see how the models can be expanded to account for such phenomena as nonlinearities in membrane I-V curves, and even for the action potential itself.

PROBLEMS

1. In the charging curve for a nerve cell membrane, given by Equation 7, how many time constants must elapse after the start of a square current stimulus before the depolarization $(E_m - E_r)$ reaches 95% of the final asymptotic value (iR_m)?

2. The charging curve in Figure 1, top part, was produced by an applied current of 9.2 nA. The time constant is 0.124 sec, and the asymptotic value of the depolarization is 6.7 mV. What is the total membrane resistance? The membrane capacitance?

3. If an *Aplysia* neuron cell body has the membrane resistance and capacitance in Problem 2 and is assumed to be a sphere 400 μm in diameter, what are the *area-specific* resistance (in $\Omega \cdot$ cm²) and capacitance (in μF/cm²)?

4. What is the *product* of the area-specific resistance and capacitance?

5. The area-specific membrane capacitance in the *Aplysia* neuron may actually be the same as found in the squid axon, about 1 μF/cm². This can be calculated as in Problem 3, using a larger area than that of a sphere (probably a correct assumption, because of the extensive infolding of the surface membrane). How many times larger is the actual area than that of a sphere?

6. What is the corrected area-specific resistance for the *Aplysia* neuron? (It is about 1,000 $\Omega \cdot$ cm² in the squid axon.)

7. Repeat Problem 1 using Equation 12 instead of Equation 7. Choose the appropriate value of the error function (erf) from the following table:

Z	erf Z	Z	erf Z
0.90	0.797	1.20	0.910
0.95	0.821	1.25	0.923
1.00	0.843	1.30	0.934
1.05	0.862	1.35	0.944
1.10	0.880	1.39	0.950
1.15	0.896	1.45	0.960

8. Show that Equation 15 reduces to Equation 16 when $\alpha = 0$. [erfc $(0) = 1$.]

REFERENCES

Adrian, R. H., and Freygang, W. H. (1962). Potassium and chloride permeability of frog muscle membrane, *J. Physiol.* **163,** 61–103.

Armstrong, C. M., and Binstock, L. (1965). Anomalous rectification in the squid giant axon injected with tetraethylammonium chloride, *J. Gen. Physiol.* **48,** 859–872.

Hodgkin, A. L. (1937). Evidence for electrical transmission in nerve, part I, *J. Physiol.* **90,** 183–210; part II, *J. Physiol.* **90,** 211–232.

Hodgkin, A. L., Huxley, A. F., and Katz, B. (1952). Measurement of current-voltage relations in the membrane of the giant axon of *Loligo, J. Physiol.* **116,** 424–448.

Hodgkin, A. L., and Rushton, W. A. H. (1946). The electrical constants of a crustacean nerve fiber, *Proc. Roy. Soc. B.* **133,** 444–479.

Kandel, E. R., and Tauc, L. (1966). Anomalous rectification in the metacerebral giant cells and its consequences for synaptic transmission, *J. Physiol.* **183,** 287–304.

Kelvin, Lord W. T. (1855). On the theory of the electric telegraph, *Proc. Roy. Soc.* **7,** 382–399.

Noble, D. (1965). Electrical properties of cardiac muscle attributable to inward going (anomalous) rectification, *J. Cell Comp. Physiol.* **66,** suppl. 2, 127–136.

Rall, W. (1959). Branching dendritic trees and motoneuron membrane resistivity, *Exptl. Neurol.* **1,** 491–527.

Rall, W. (1960). Membrane potential transients and membrane time constant of motoneurons, *Exptl. Neurol.* **2,** 503–532.

4

IONIC PROPERTIES OF RESTING
AND ACTIVE MEMBRANES

ASYMMETRIC DISTRIBUTION OF IONS
ACROSS THE MEMBRANE

The electrical analogues discussed in Chapter 3 provide a convenient description of the responses of real excitable membranes to applied stimuli. However, these models give no indication of the ions which are carrying the currents, or of the molecular mechanisms of ion transport involved. As mentioned in Chapter 1, Bernstein (1902, 1912) had developed a theory of membrane potential based on the asymmetric distribution of ions across the cell wall. This distribution is illustrated for squid axoplasm and blood in Table 1. It can be seen that the potassium concentration is much higher inside the axon than outside and that the

Table 1. Approximate concentrations of ions in axoplasm and blood of squid (after Hodgkin, 1951, based on specific gravity of axoplasm and blood of 1.025 g/ml).

	Axoplasm, mmol/l	Blood, mmol/l
K	397	20
Na	50	437
Cl	40	556
Ca	0.4	10
Mg	10	54

sodium and chloride concentrations are much higher outside. This situation could theoretically arise by at least two mechanisms: (1) There is in the axoplasm a considerable amount of negatively charged protein which cannot cross the membrane, and the resting membrane is mainly permeable to potassium and chloride. Thus, the K^+ and Cl^- could distribute in the directions shown in Table 1 according to a DONNAN EQUILIBRIUM. This mechanism is outlined in Figure 1. In this system, a membrane permeable to K^+ and Cl^- separates one compartment containing K^+ and Cl^- from another containing K^+, Cl^-, and an impermeant anion, A^-. For electroneutrality to exist in the outside solution,

$$[K]_o = [Cl]_o \tag{1}$$

and in the inside

$$[K]_i = [Cl]_i + [A] \tag{2}$$

Potassium has a tendency to move out of this membrane because it is more concentrated inside. But each K^+ ion which moves out must be accompanied by a Cl^- ion, in order to maintain electroneutrality. This leads to an accumulation of Cl^- outside the membrane, which creates a concentration gradient opposite to that for K^+. These driving forces become equal at equilibrium, when the concentration ratios for the diffusible ions are equal; i.e.,

$$\frac{[K]_o}{[K]_i} = \frac{[Cl]_i}{[Cl]_o} \tag{3}$$

Taking $[Cl]_i$ from Equation 3 and substituting into Equation 2 gives

$$[K]_i = \frac{[K]_o}{[K]_i}[Cl]_o + [A] \tag{4}$$

Taking $[Cl]_o$ from Equation 1 and substituting into Equation 4 gives

$$[K]_i = \frac{[K]_o^2}{[K]_i} + [A] \tag{5}$$

or

$$[K]_i^2 = [K]_o^2 + [A][K]_i \tag{6}$$

So

$$[K]_i > [K]_o \quad \text{and} \quad [Cl]_o > [Cl]_i \tag{7}$$

Equation 3 does not fit exactly with the data for K^+ and Cl^- in Table 1, but it does predict the proper directions of the change of K^+ and Cl^- concentration from inside to outside. The K^+ and Cl^- concentrations inside and outside of frog muscle fibers follow a Donnan rela-

INSIDE	OUTSIDE	
K$^+$	K$^+$	Asymmetric ion distribution resulting from a Donnan equilibrium. A$^-$, impermeant anion.
Cl$^-$	Cl$^-$	
A$^-$		

1

tionship quite closely. However, in this case an excess of impermeant sodium ions on the outside maintains the osmotic equilibrium (Boyle and Conway, 1941).

(2) The second mechanism which operates in excitable cells to accumulate potassium ions also acts to reduce the internal sodium concentration; this is the METABOLIC PUMP. As shown by Hodgkin and Keynes (1955), the squid axon membrane is continually extruding sodium ions from inside to outside by a process which requires adenosine triphosphate as an energy source. In addition, this sodium pump is coupled to the entry of potassium ions. If the external potassium is removed, the outward pumping of sodium is greatly reduced. One potassium ion must enter for each two or three sodium ions which are pumped out (Caldwell *et al.*, 1960). Thus, the pump naturally functions to increase internal K$^+$ and reduce internal Na$^+$. As will be discussed in Chapter 9, this type of pump operates in all known excitable cells, although the Na–K coupling ratio may vary considerably from one system to another.

THE NERNST EQUATION

If one is given the asymmetric distribution of Na, K, and Cl$^-$ ions across excitable cell membranes, the resting transmembrane potentials may be derived from the ionic theory. Much of this theory is due to Nernst, who developed it for nonliving systems. Thus, although Bernstein first applied this theory to biological membranes, the equation for transmembrane potential in a single-electrolyte system is still called the Nernst equation. It may be derived easily for a system in which, for instance, the K$^+$ and Cl$^-$ concentrations are different on two sides of a membrane: First, the electric potential is assumed to vary throughout the membrane. Second, because the system is *not* in equilibrium, ion FLUXES will be present across the membrane. These fluxes each consist of two terms: one resulting from diffusional forces and the other from the effect of the membrane electric field on the charged ions. Thus, for the flux of potassium

$$J_K = -D_K\left(\frac{d[K]}{dx} + [K]\frac{F}{RT}\frac{d\psi}{dx}\right)$$

(8)

where J_K = flux of K$^+$ in mol/sec/cm^2 of membrane
$\quad\quad D_K$ = diffusion constant for K$^+$
$\quad\quad \psi$ = potential at any point
$\quad\quad F$ = Faraday constant
$\quad\quad R$ = universal gas constant
$\quad\quad T$ = temperature, °K
Similarly, for the flux of chloride

$$J_{Cl} = -D_{Cl}\left(\frac{d[Cl]}{dx} - [Cl]\frac{F}{RT}\frac{d\psi}{dx}\right) \tag{9}$$

With no net current imposed across the membrane, the currents due to K$^+$ and Cl$^-$ should sum to zero, that is,

$$I = Z_K F J_K + Z_{Cl} F J_{Cl} = 0 \tag{10}$$

where I = total membrane current
$\quad\quad Z$ = valence of ion
$\quad\quad F$ = Faraday constant
So, substituting Equations 8 and 9 into Equation 10 gives

$$-D_K\frac{d[K]}{dx} + D_{Cl}\frac{d[Cl]}{dx} - \frac{F}{RT}\frac{d\psi}{dx}\left(D_K[K] + D_{Cl}[Cl]\right) = 0 \tag{11}$$

or

$$-\frac{F}{RT}\frac{d\psi}{dx} = \frac{D_K\dfrac{d[K]}{dx} - D_{Cl}\dfrac{d[Cl]}{dx}}{D_K[K] + D_{Cl}[Cl]} \tag{12}$$

A simplifying assumption is now made, that for electroneutrality, $[K] = [Cl]$ at all points. Then

$$-\frac{F}{RT}\frac{d\psi}{dx} = \frac{(D_K - D_{Cl})\dfrac{d[K]}{dx}}{(D_K + D_{Cl})[K]} \tag{13}$$

This may be integrated directly as

$$\frac{F}{RT}(\psi_o - \psi_i) = \frac{D_K - D_{Cl}}{D_K + D_{Cl}}\ln\frac{[K]_o}{[K]_i} \tag{14}$$

If we now introduce the definition of MOBILITY as

$$u = \frac{D_K}{RT} \quad\quad v = \frac{D_{Cl}}{RT} \tag{15}$$

then Equation 14 becomes

$$E = \frac{u - v}{u + v}\frac{RT}{F}\ln\frac{[K]_o}{[K]_i} \tag{16}$$

where $E = \psi_o - \psi_i$ = transmembrane potential
$[K]_o$ = potassium concentration on outside
$[K]_i$ = potassium concentration on inside
This is the same as Equation 1, chapter 1, due to Bernstein.

If the membrane is now assumed to be permeable only to potassium, i.e., having $v = 0$, then

$$E = \frac{RT}{F} \ln \frac{[K]_o}{[K]_i} \qquad (17)$$

This expression for the potential across a membrane which is permeable only to one ionic species is often called the NERNST EQUATION. Because RT/F is about 25 mV at room temperature and $\ln([K]_o/[K]_i) \approx 2.3 \log([K]_o/[K]_i)$, Equation 17 predicts about a 58-mV change in potential for a ten-fold increase in $[K]_o$.

THE GOLDMAN-HODGKIN-KATZ EQUATION

Bernstein's theory could account very well for the variation of resting potential with external potassium concentration, but it could not explain the inside-positive action potential observed by Hodgkin and Huxley (1939). The first step in modifying the ionic theory to embody this result was Goldman's work (1943), in which he related the membrane potential to the concentrations of several different electrolytes in the membrane itself. An assumption in his model was that the electric field, or rate of change of potential with distance, in the membrane was a constant. Hence, his treatment is often referred to as CONSTANT-FIELD THEORY. This was used by Hodgkin and Katz (1949) to account for the membrane potential of the squid axon. These authors performed many experiments in which they varied the external Na and K concentrations around the axons. They concluded, in agreement with Bernstein's theory, that the resting membrane was selectively permeable to potassium ions. However, in the active condition the squid axon membrane became selectively permeable mainly to sodium ions and not to all ions, as Bernstein had thought. This was reflected in the fact that the action potential overshoot varied only with external sodium concentration and was insensitive to potassium concentration. The modified equation used to describe the axon membrane potentials was

$$E = \frac{RT}{F} \ln \frac{P_K[K]_o + P_{Na}[Na]_o + P_{Cl}[Cl]_i}{P_K[K]_i + P_{Na}[Na]_i + P_{Cl}[Cl]_o} \qquad (18)$$

where

$[K]_o$, $[Na]_o$, and $[Cl]_o$ are the concentrations of these ions outside the membrane

[K]$_i$, [Na]$_i$, and [Cl]$_i$ are the concentrations in the axoplasm

P_K, P_{Na}, and P_{Cl} are the permeabilities for each ion

Equation 18 is now conventionally referred to as the GOLDMAN-HODGKIN-KATZ EQUATION.

By using Steinbach and Spiegelman's (1943) data for internal ion concentrations, Hodgkin and Katz found that the variation of the resting potential with external potassium concentration could be accounted for if they assumed $P_K:P_{Na}:P_{Cl} = 1:0.04:0.45$. Furthermore, they could account for the internal-positive action potential quantitatively by assuming a 500-fold increase in sodium permeability in the active state to make $P_K:P_{Na}:P_{Cl} = 1:20:0.45$. This was the first demonstration that the SODIUM HYPOTHESIS, or the idea of a selective increase in the sodium permeability, could explain the reversal of transmembrane potential during the nerve impulse.

The following year the sodium hypothesis was tested in muscle fibers (Nastuk and Hodgkin, 1950). Boyle and Conway (1941) had shown that the intracellular sodium concentration in frog muscle fibers was about one-eighth of that in the external fluid; so an increase in sodium permeability could give rise to the action potential. Nastuk and Hodgkin replaced the sodium outside the muscle fiber with choline, an organic cation with the following structure:

$$CH_3-\overset{\displaystyle CH_3}{\underset{\displaystyle CH_3}{N^+}}-CH_2-CH_2OH$$

They found that the active membrane potential varied 58 mV for each ten-fold change in external sodium concentration. The active muscle fiber membrane evidently behaved like a sodium electrode. This behavior was contrasted with that of the resting membrane, which was considered to be permeable mainly to potassium and chloride ions.

MEASUREMENT OF P_{Na}/P_K RATIO

It is possible to estimate the P_{Na}/P_K ratio in some excitable cells by using a simplified version of Equation 18. The term $P_{Na}[Na]_i$ is neglected as being much smaller than the terms to which it is added. $P_{Cl}[Cl]_i$ is neglected for a reason which will be indicated, and the experiments are conducted in Cl-free solutions. Equation 18 then becomes

$$E = \frac{RT}{F} \ln \frac{[K]_o + P_{Na}/P_K[Na]_o}{[K]_i} \qquad (19)$$

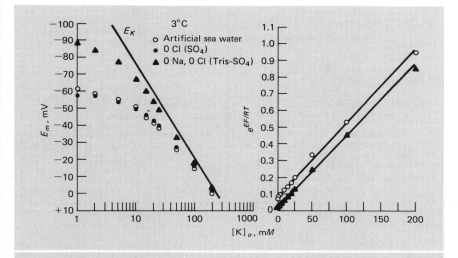

Left: Variation of membrane potential in a molluscan ganglion cell with external K concentration in normal, Cl-free, and Na-free solutions. Solid line indicates 58 mV per tenfold change in $[K]_o$. Right: Variation of $e^{EF/RT}$ with external K concentration. (Gorman and Marmor, 1970.)

and

$$e^{EF/RT} = \frac{[K]_o}{[K]_i} + \frac{P_{Na}/P_K[Na]_o}{[K]_i}$$ (20)

In Figure 2, these equations are plotted against the external potassium concentration and compared with the real membrane potential of a molluscan ganglion cell (from Gorman and Marmor, 1970). On the left side, the curve of potential versus $[K]_o$ in chloride-free solution (solid dots) is about the same as that in chloride-containing solution (circles); so the resting membrane is relatively impermeable to chloride, justifying the dropping of $P_{Cl}[Cl]_i$ in Equation 18. Replacement of external sodium with tris (tris-hydroxymethyl aminomethane, an impermeant cation) (triangles) causes the potential to follow the 58-mV line expected for a potassium-selective membrane much more closely. On the right side of the figure, the values of $e^{EF/RT}$ agree well with Equation 20. When external sodium is replaced (triangles), the intercept changes and the slope is unaffected. The slope is equal to $1/[K]_i$, and the change in intercept is $P_{Na}[Na]_o/P_K[K]_i$. If one knows $1/[K]_i$ and $[Na]_o$, it is possible to calculate P_{Na}/P_K. Alternatively, the potassium concentration may be held constant and $[Na]_o$ varied. In this case, the slope of the resulting line is $P_{Na}/P_K[K]_i$. The importance of the P_{Na}/P_K ratio may be emphasized by noting that the different ratios for the resting and active states of excitable membranes is crucial for the production of action potentials.

ALKALI-CATION SELECTIVITY

A further description of the membrane in a particular state is given by its alkali-cation selectivity, or its relative permeabilities to Li^+, Na^+, K^+, Rb^+, and Cs^+. This may be obtained using a modified version of Equation 18: If a cation I^+ is the only external cation and chloride-free solutions are used, then the membrane potential becomes

$$E = \frac{RT}{F} \ln \frac{P_I/P_K[I]_o + P_{Cl}/P_K[Cl]_i}{[K]_i} \tag{21}$$

and

$$e^{EF/RT} = \frac{P_I/P_K[I]_o}{[K]_i} + \frac{P_{Cl}/P_K[Cl]_i}{[K]_i} \tag{22}$$

The internal chloride term is included because we have no *a priori* knowledge of how large it is. If the concentration of each cation is systematically changed, one at a time, then the graphs of $e^{EF/RT}$ versus $[I]_o$ should have slopes equal to $P_I/P_K[K]_i$. The value of $[K]_i$ may be obtained from the slope of the line when I^+ is K^+. Thus the relative permeabilities P_I/P_K may be found for all five cations.

Without any restricting assumptions, there are $5! = 120$ possible sequences of the five permeability ratios. Interestingly, only 11 of

Table 2. Observed sequences of alkali-cation selectivities (from Eisenman, 1961).

1.	$Cs^+ > Rb^+ > K^+ > Na^+ > Li^+$
2.	$Rb^+ > Cs^+ > K^+ > Na^+ > Li^+$
3.	$Rb^+ > K^+ > Cs^+ > Na^+ > Li^+$
4.	$K^+ > Rb^+ > Cs^+ > Na^+ > Li^+$
5.	$K^+ > Rb^+ > Na^+ > Cs^+ > Li^+$
6.	$K^+ > Na^+ > Rb^+ > Cs^+ > Li^+$
7.	$Na^+ > K^+ > Rb^+ > Cs^+ > Li^+$
8.	$Na^+ > K^+ > Rb^+ > Li^+ > Cs^+$
9.	$Na^+ > K^+ > Li^+ > Rb^+ > Cs^+$
10.	$Na^+ > Li^+ > K^+ > Rb^+ > Cs^+$
11.	$Li^+ > Na^+ > K^+ > Rb^+ > Cs^+$

these sequences are observed in living or nonliving systems which bind cations. They are shown in Table 2. Sequence 1 is the lyotropic series, arranged in order of increasing hydrated size. Sequence 11 is arranged in order of increasing nonhydrated size. In the squid giant axon, sequence 4 describes the ionic dependency of the resting potential, and sequence 11 applies to the peak of action potential (Eisenman, 1965). The other sequences are seen in various types of glass electrodes and in frog skin, frog and lobster muscle, *E. coli,* yeasts, red blood cells, *Chlorella,* crab nerve, and some enzymes (Eisenman, 1965).

EISENMAN THEORY

A simple, unifying theory which can explain the occurrence of only 11 out of 120 possible sequences was put forth by Eisenman (1961). This theory postulates that differences in *anionic site strength* in the membrane determine which of the 11 sequences is preferred in a given system. The argument goes as follows: In order for a cation in solution, I^+, to bind to a membrane site, S^-, it must first detach itself from the associated water molecule or molecules, W. The overall reaction is

$$I^+W + S^- \rightleftharpoons I^+S^- + W \qquad (23)$$

The standard free energy change for this reaction is

$$\Delta G_i^* = \Delta G_{\text{ion-site}} - \Delta G_{\text{ion-water}} \qquad (24)$$

At this point it is assumed that the most significant differences between $\Delta G_{\text{ion-site}}$ and $\Delta G_{\text{ion-water}}$ lie in the different electrostatic binding energies for the site and the water molecules. For a monovalent site, the binding energy is found from Coulomb's law as

$$\Delta U_{\text{ion-site}} = \frac{332q^+q^-}{r^+ + r^-} \qquad (25)$$

where

$\Delta U_{\text{ion-site}}$ = binding energy, kcal/mol
q^+ = charge of the cation, electronic charges
q^- = charge of the anionic site, electronic charges
r^+ = crystal radius of the cation, Å
r^- = crystal radius of the anionic site, Å

The site strength may thus be varied by changing q^- or r^-. A smaller site permits a closer approach by a cation, which is thus more strongly bound. If, for instance, the radius of the anionic site is varied, then the binding energy $\Delta U_{\text{ion-site}}$ may be calculated for several different radii by using Equation 25. Values of the crystal radii and hydration energies for the alkali cations are listed in Table 3. We may thus calculate

Table 3. Hydration energies and Goldschmidt crystal radii for the alkali cations. Hydration energies based on coulombic forces between single Rawlinson-type water molecules and each cation (see Eisenman, 1961).

	r^+, Å	$\Delta U_{\text{ion-water}}$, kcal/mol
Li$^+$	0.78	-24.2
Na$^+$	0.98	-20.5
K$^+$	1.33	-15.8
Rb$^+$	1.49	-14.1
Cs$^+$	1.65	-12.7

the overall change in binding energy for dehydration plus site binding as

$$\Delta U^* = \Delta U_{\text{ion-site}} - \Delta U_{\text{ion-water}} \qquad (26)$$

This quantity actually gives the ion selectivity for each radius of the anionic site, because alkali cations with the most negative values of ΔU^* will be preferred. These are usually calculated with respect to a certain cation such as Cs$^+$; for instance:

$$\Delta U^R = \Delta U_{\text{ion}}^* - \Delta U_{\text{Cs}}^* \qquad (27)$$

A sample calculation may elucidate this method: When $r^- = 3.5$Å,

	Li$^+$	Na$^+$	K$^+$	Rb$^+$	Cs$^+$
$r^+ + r^-$, Å	4.28	4.48	4.83	4.99	5.15
$\Delta U_{\text{ion-site}}$, kcal/mol	-77.6	-74.1	-68.7	-66.5	-64.5
$\Delta U_{\text{ion-water}}$, kcal/mol	-24.2	-20.5	-15.8	-14.1	-12.7
ΔU^*, kcal/mol	-53.4	-53.6	-52.9	-52.4	-51.8
ΔU^R, kcal/mol	-1.6	-1.8	-1.1	-0.6	0

This is sequence 10, since the order of ion selectivities is Na > Li > K > Rb > Cs.

As the anionic site radius, r^-, is varied, different curves of ΔU^R versus r^- are produced for each cation, as shown in Figure 3. The size of ΔU^R represents the selectivity of the system (including ion-site binding and dehydration) for a particular ion as a function of anionic site radius. Because all the selectivities are referred to Cs$^+$, the ΔU^R for Cs$^+$ is invariant. However, as r^- is varied, the system selectivity

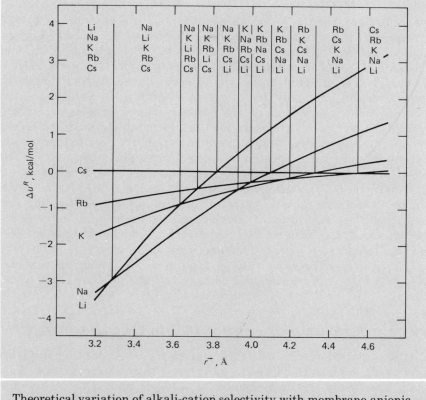

3

Theoretical variation of alkali-cation selectivity with membrane anionic site radius, r^-. ΔU^R, difference in energy of site binding and dehydration relative to that for cesium. (Same method as Eisenman, 1961.)

changes through the *11 known sequences*. At low anionic site radii (high site strength), Li^+ is preferred. A similar set of curves is obtained if the value of q^- is varied. Differences in anionic site strength could arise from inductive effects in membrane molecules (changing q^-) or from coordination effects, where several anionic sites are close enough to each other to interact with single cations.

Membrane selectivities for the halides, F^-, Cl^-, I^-, and Br^-, and for the divalent cations Sr^{++}, Ba^{++}, Ca^{++}, and Mg^{++} have also been calculated according to this theory and related to observed sequences (for review see Diamond and Wright, 1969). This powerful approach has explained many of the confusing properties of membranes, and it must be counted as a significant achievement of biophysical theory.

CALCIUM EFFECTS

The effects of calcium on nerve membranes were not covered by the Hodgkin-Katz paper in 1949, although some work had been done on the subject by that time (Arvanitaki, 1939; Brink *et al.*, 1946; and Hodgkin, 1948). The usual result of lowering external calcium concentration in frog or squid nerve was to lower the critical, or threshold, depolarization for production of action potentials. In some cases, the neurons fired spontaneously in Ca-free solutions. Prolonged application of Ca-free solutions could lead to irreversible damage, a fall in membrane resistance, and subsequent inability to produce action potentials even with repeated washing in normal saline solution. The effect of calcium on nerve membranes was thus considered to be one of "stabilization," or prevention of spontaneous firing due to small fluctuations in membrane potential, and maintenance of membrane integrity. The importance of external calcium for keeping nerve membranes intact has been studied more recently by Oomura *et al.* (1961) and by Tasaki *et al.* (1967), among others.

Calcium can also substitute for sodium as the inward-current carrier in a variety of nerve and muscle cells, but that story is for Chapter 8.

PARALLEL-CONDUCTANCE MODEL OF THE MEMBRANE

The Goldman-Hodgkin-Katz theory gives a good accounting of the behavior of membranes bathed in solutions of several electrolytes. However, this theory should be applied to only *one* state of the membrane (resting or active) at a time because it does not accurately explain the transition from one state to another. That is, the process of excitation of nerve and muscle is now thought to result from the sequential activation of independent ionic channels, each with its own particular selectivities, as is discussed in Chapter 7.

A more representative description of the change from inactive to active membrane is provided by the PARALLEL-CONDUCTANCE MODEL described by Hodgkin and Huxley (1952a) and further elaborated by Hodgkin and Horowicz (1959). Here, the membrane is thought to contain three parallel channels, one for potassium ions, one for sodium ions, and one for chloride. Each channel is described by a battery, E, in series with a conductance, g (equal to the inverse of the resistance of the channel). The membrane capacitance, C_m, is across the whole ensemble, as shown in Figure 4. This model does not include longitudinal spread of current. Hence, it is only applicable to (1) small areas of membrane, such as isolated nodes of Ranvier, (2) axons with long

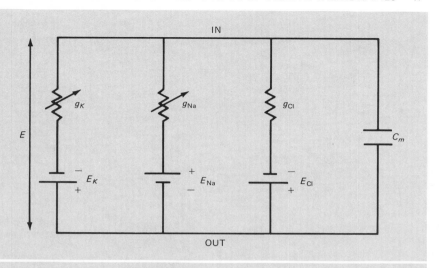

4

Parallel-conductance model of nerve membrane, for which one assumes independent conductance channels for K, Na, and Cl.

metal electrodes inserted into them to make the inside isopotential, or (3) roughly spherical nerve cell bodies, where the inside is isopotential.

When $dE/dt = 0$, at rest, at the trough of hyperpolarization, or at the peak of the action potential, then the capacitive current ($C_m \, dE/dt$) disappears, and the membrane potential is given by

$$E = \frac{E_K g_K + E_{Na} g_{Na} + E_{Cl} g_{Cl}}{g_K + g_{Na} + g_{Cl}} \tag{28}$$

where $E_I = \dfrac{RT}{F} \ln \dfrac{[I]_o}{[I]_i}$ is the equilibrium potential for each ion.

This is sometimes written

$$E = E_K T_K + E_{Na} T_{Na} + E_{Cl} T_{Cl} \tag{29}$$

where

$$T_I = \frac{g_I}{g_K + g_{Na} + g_{Cl}} \tag{30}$$

T_I is called the *transport number* for the ion. In this formulation, T_{Na} is small compared with T_K in the resting membrane, and it increases during activity. This model was used by Hodgkin and Huxley in recon-

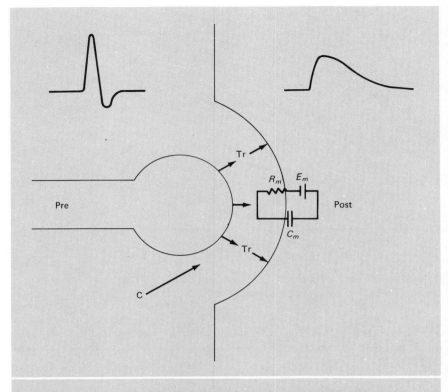

5

Sequence of events in chemical transmission. Circuit diagram shows electrical model of postsynaptic membrane.

structing the squid action potential from theoretically obtained conductances, as will be discussed in Chapters 5 and 6.

SYNAPTIC POTENTIALS

The above parallel-conductance model is also useful in characterizing the responses of nerve and muscle cells to synaptic stimulation. A SYNAPSE is the junction across which one nerve cell excites another, or across which a nerve excites a muscle or gland cell. In some cases, this is accomplished by movement of a chemical transmitter substance from one cell to the next. In others, the presynaptic and postsynaptic cells may be electrically coupled. The sequence of events in chemical transmission is outlined in Figure 5. An action potential invades the terminal, or the presynaptic neuron (Pre), causing the release of transmitter (Tr) into the synaptic CLEFT (C). (The cleft, or gap between the cells, is much more narrow than shown, usually on the order of 150Å.) A short time after the arrival of the presynaptic impulse (about 0.5 msec in mammalian neurons, longer in invertebrates), the transmitter diffuses across the cleft and acts on the membrane of the postsynaptic cell (Post).

In the case shown, the effect of the transmitter is a depolarization. This type of response is called an EXCITATORY POSTSYNAPTIC POTENTIAL (EPSP). The EPSP is the unit of excitatory synaptic transmission. If large enough (suprathreshold), it can produce an action potential in the postsynaptic cell. If several *sub*threshold EPSPs occur within a short time, they can add together to raise the membrane potential above threshold, and produce postsynaptic action potentials. In other types of synapses, the transmitter acts to hyperpolarize the postsynaptic cell. The response is then called an INHIBITORY POSTSYNAPTIC POTENTIAL (IPSP). The IPSP acts to move the membrane potential away from threshold. It is the unit of synaptic inhibition, where activity in one cell causes less activity in the next.

One important question for membrane biophysicists is the mechanism by which the synaptic transmitter produces an EPSP or IPSP in the postsynaptic cell. In this regard, the parallel-conductance view of the postsynaptic cell, as shown in Figure 6, is useful. E_r and g_r are, respectively, the resting potential and conductance of the postsynaptic

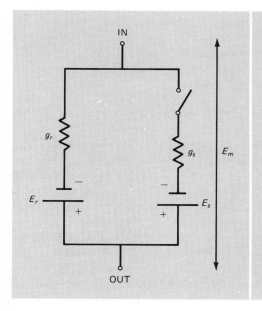

Electrical analogue of postsynaptic membrane. Discussion of the figure may be found in the text.

6

membrane. (The capacitance has been ignored for simplicity; the following discussion will deal with steady potentials toward which the membrane moves during a postsynaptic response.) E_s and g_s are the corresponding properties of synaptic ionic channels through the membrane. Thus, the transmitter is regarded as activating the synaptic channels, or closing the switch in Figure 6. E_m then changes from E_r to a value somewhere between E_r and E_s. The value of E_m during an EPSP or IPSP may be calculated from the model as follows: With no outside current applied during the synaptic event, the current in the resting channel, I_r, and that in the synaptic channel, I_s, must sum to zero:

$$I_r + I_s = 0 \tag{31}$$

where

$$\begin{aligned} I_r &= g_r(E_m - E_r) \\ I_s &= g_s(E_m - E_s) \end{aligned} \tag{32}$$

From Equations 31 and 32, the final value of E_m is

$$E_m = \frac{E_r g_r + E_s g_s}{g_r + g_s} \tag{33}$$

For an EPSP, for instance, E_s is more positive than E_r; so E_m moves in the depolarizing direction. (Although this shift in potential is sudden in the model, in real cells it is gradual because of the smoothing action of the membrane capacitance and the time course of transmitter release.) The falling phase of the postsynaptic potential is initiated when the transmitter action stops, and the switch is opened in Figure 6. The *change* in membrane potential from resting which is produced by an EPSP or IPSP may be calculated as

$$E_m - E_r = \frac{g_s(E_s - E_r)}{g_r + g_s} \tag{34}$$

The physical meaning of this model is that during a synaptic event an additional conductance is added to that already present in the resting membrane. The synaptic current, I_s, is carried by ions moving through the transmitter-activated channels. This type of ionic model of the postsynaptic membrane was first used by Fatt and Katz (1951) to describe the properties of the frog neuromuscular junction.

From Equation 34, the synaptic event will have the form of an EPSP if E_s is less negative than E_r and of an IPSP if E_s is more negative. If the prestimulus level of potential is made more negative with artificially applied inward current, then an EPSP will be increased in size, and an IPSP will be reduced. In fact, it is possible to *reverse* the direction of an IPSP, as shown in Figure 7. These records were ob-

tained from a spinal motoneuron by stimulation of an inhibitory afferent nerve (Coombs *et al.*, 1955). The normal resting potential was
-74 mV, and the IPSP size with no applied current is shown in D. In
A to C, a depolarizing (outward) current was injected into the motoneuron through one side of a double-barreled microelectrode, causing
the IPSP amplitude to increase. In E to G, hyperpolarizing (inward)
current was applied, causing the IPSP to reverse near -82 mV. This
behavior can be understood in terms of the model by assuming that
the equilibrium potential, E_s, is about -82 mV and that when the post-

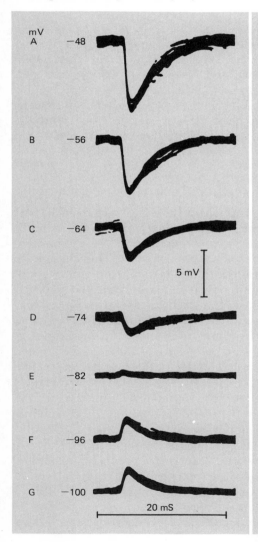

Reversal of inhibitory postsynaptic potential in a cat
motoneuron by means of injected current. Outward current applied in A to C; no
current in D; inward cur-
rent in E to G. Values on
left show potential at which
membrane was held before
synaptic stimulation.
(Coombs *et al.*, 1955.)

7

synaptic conductance increases the membrane potential is drawn toward E_s. If the potential is held below (more negative than) E_s prior to stimulation of the afferent nerve, then the IPSP turns over.

In addition, Coombs *et al.* were able to change the sign of the IPSP by altering ion concentrations inside the motoneuron: injection of Cl^- ions reversed the IPSP at normal membrane potentials. The effect of the injection on the model parameters was simply to make E_s more positive. Thus, E_s followed E_{Cl}, the Nernst potential for chloride, indicating that the inhibitory synaptic channel included a large amount of chloride conductance.

The value of E_s in the model can also be changed by altering *external* ion concentrations, which has been studied in a variety of preparations (Trautwein and Dudel, 1958; Furshpan and Potter, 1959; Hagiwara *et al.*, 1960; Furukawa and Furshpan, 1963; Kerkut and Thomas, 1964; and Gerschenfeld and Chiarandini, 1964 — for review see Ginsborg, 1967). In these and other cases, the parallel-conductance model is able to account for such properties of postsynaptic potentials as variation with prestimulus level of potential and with external and internal ion concentrations.

GENERATOR POTENTIALS

Another area of neurophysiology in which the conductance model has been widely applied is that of receptor physiology: in general, stimuli which impinge on an organism produce changes in the membrane potential of the appropriate SENSORY RECEPTOR cells. This change in potential (usually a depolarization) either sets up a train of nerve impulses in the receptor cell itself or depolarizes a connected neuron, initiating the train. In either case, the model is useful in describing the mechanism by which a stimulus brings about a change in the receptor membrane potential. In Figure 8 is shown an intracellular recording from the crayfish stretch receptor cell during mechanical displacement. A slight stretch was applied at the first arrow, which produced an irregular discharge. Increasing the stretch (second arrow) caused a corresponding increase in firing frequency. After the tension was released (downward arrow), the membrane potential returned to the resting level. If a slightly smaller stretch was applied, the resulting depolarization could be seen without action potentials. This depolarization, or GENERATOR POTENTIAL, causes the production of action potentials in the same way as direct intracellular stimulation: it simply raises the membrane potential above the threshold for spiking. Edwards *et al.* (1963) showed that the depolarizing response to stretch was chiefly dependent on the external Na^+ concentration, and less so

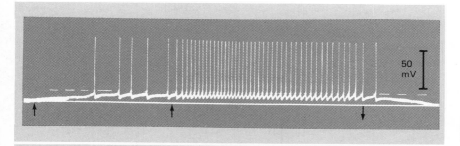

8

Depolarizing response to stretching of crayfish stretch receptor. Slight stretch applied and maintained at first arrow. Stretch increased at second arrow; released at third arrow. Broken line indicates threshold for production of action potentials. (Eyzaguirre and Kuffler, 1955.)

on K^+ and Ca^{++}. This behavior can be accounted for by a model such as that in Figure 6, if E_s is considered to be near the sodium equilibrium potential and g_s is a conductance activated by mechanical stretching of the membrane. Recent studies have confirmed the earlier interpretations of the conversion of mechanical displacement into impulse trains in this receptor (Loewenstein et al., 1963; and Nakajima and Onadera, 1969).

Generator potentials have also been recorded in the pacinian corpuscle (Diamond et al., 1958), in the photoreceptors of the horseshoe crab (Fuortes, 1959) and barnacle (Koike, et al., 1971), and in taste receptor cells (Kimura and Beidler, 1961). In the taste receptor, it is probably more appropriate to call the depolarizing response a RE-CEPTOR POTENTIAL, because no action potentials are produced in these epithelially derived cells. This depolarization is thought to produce a generator potential in the connected nerve terminals, which gives rise to the train of impulses.

All the above receptor responses, like the synaptic responses mentioned earlier, can be characterized by the parallel-conductance model. While not a complete description of either receptor or synaptic membranes, it has two significant advantages: (1) It is simple, and (2) it provides a unifying framework for the experiments which advance our knowledge of these systems.

PROBLEMS

1. Calculate E_K, the potassium equilibrium potential, for the data in Table 1 using Equation 17.

2. Similarly, calculate the sodium and chloride equilibrium potentials in Table 1. Which of the three equilibrium potentials you have calculated are closest to each other?

3. Brown *et al.* (1970) calculated the following constants for the *Aplysia* giant cell:

$$[Na]_o = 337 \text{ m}M$$
$$[Na]_i = 50 \text{ m}M$$
$$[K]_o = 6 \text{ m}M$$
$$[K]_i = 168 \text{ m}M$$
$$[Cl]_o = 340 \text{ m}M$$
$$[Cl]_i = 41 \text{ m}M$$

and for the resting cell membrane

$$P_K : P_{Na} : P_{Cl} = 1.0 : 0.019 : 0.381$$

What is the resting potential predicted by the Goldman-Hodgkin-Katz equation? [Note that $(RT/F) \ln A = 58 \log A$.]

4. In Problem 3, what would be the effect on resting potential of a ten-fold increase in external potassium concentration?

5. In the parallel-conductance model of the above cell, the relevant values for the resting membrane are:

$$E_K = -84 \text{ mV}$$
$$E_{Na} = +48 \text{ mV}$$
$$E_{Cl} = -53 \text{ mV}$$
$$g_K = 0.57 \ \mu\text{mho}$$
$$g_{Na} = 0.11 \ \mu\text{mho}$$
$$g_{Cl} = 0.32 \ \mu\text{mho}$$

What is the predicted resting potential?

6. In Problem 5, to what value must g_{Na} increase to make the membrane potential $+40$ mV if all other quantities remain constant?

7. For the data in Figure 2, right side, the slopes of the lines of exp (EF/RT) versus $[K]_o$ are both equal to 0.00426. From Equation 20, what is the predicted value of $[K]_i$?

8. In Problem 7, the external sodium concentration is 480 mM, and the change in intercept on replacement of external sodium is 0.057. What is the P_{Na}/P_K ratio?

9. Continue the calculation of selectivity sequences illustrated in the text, as the radius of the anionic site is increased. Try values of r^- which lie in the range shown in Figure 3, and see if you can reproduce the values of ΔU^R in that figure.

10. The values in Figure 6 appropriate to the frog neuromuscular junction are $E_r = -90$ mV, $g_r = 5 \times 10^{-6}$ mho, $E_s = -15$ mV, and $g_s = 5 \times 10^{-5}$ mho. Calculate the peak amplitude of the end-plate potential using the parallel-conductance model of the postsynaptic membrane.

REFERENCES

Arvanitaki, A. (1939). Recherches sur la réponse oscillatoire locale de l'axone géant isolé de «Sepia», *Arch Int Physiol.* **49**, 209–256.

Bernstein, J. (1902). Untersuchungen zur Thermodynamik der bioelektrischen Ströme, *Pflüger's Arch Ges. Physiol.* **92**, 521–562.

Bernstein, J. (1912). *Elektrobiologie,* Braunschweig, Vieweg u. Sohn.

Boyle, P. J., and Conway, E. J. (1941). Potassium accumulation in muscle and associated changes, *J. Physiol.* **100**, 1–63.

Brink, F., Bronk, D. W., and Larrabee, M. G. (1946). Chemical excitation of nerve, *Ann. N.Y. Acad. Sci.* **47**, 457–485.

Brown, A. M., Walker, J. L., and Sutton, R. B. (1970). Increased chloride conductance as the proximate cause of hydrogen ion concentration effects in *Aplysia* neurons, *J. Gen. Physiol.* **56**, 559–582.

Caldwell, P. C., Hodgkin, A. L., Keynes, R. D., and Shaw, T. I. (1960). The effects of injecting "energy-rich" phosphate compounds on the active transport of ions in the giant axons of *Loligo, J Physiol.* **152**, 561–590.

Coombs, J. S., Eccles, J. C., and Fatt, P. (1955). The specific ionic conductances and the ionic movements across the motoneuronal membrane that produce the inhibitory postsynaptic potential, *J. Physiol.* **130**, 326–373.

Diamond, J., Gray, J. A. B., and Inman, D. R. (1958). The relation between receptor potentials and the concentration of sodium ions, *J. Physiol.* **142**, 382–394.

Diamond, J. M., and Wright, E. M. (1969). Biological membranes: the physical basis of ion and nonelectrolyte selectivity, *Ann. Rev. Physiol.* **31**, 581–646.

Edwards, C., Terzuolo, C. A., and Washizu, Y. (1963). The effect of changes of the ionic environment upon an isolated crustacean sensory neuron, *J. Neurophysiol.* **26**, 948–957.

Eisenman, G. (1961). On the elementary atomic origin of equilibrium ionic specificity, *Symposium on Membrane Transport and Metabolism,* A. Kleinzeller and A. Kotyk (ed.), New York, Academic, 163–179.

Eisenman, G. (1965). Some elementary factors involved in specific ion permeation, *Proc 23rd Int Congr. Physiol. Sci , Tokyo,* Amsterdam, Excerpta Med. Found., 489–506.

Eyzaguirre, C., and Kuffler, S. W. (1955). Processes of excitation in the dendrites and in the soma of single isolated sensory nerve cells of the lobster and crayfish, *J. Gen. Physiol.* **39**, 87–119.

Fatt, P., and Katz, B. (1951). An analysis of the end-plate potential recorded with an intracellular electrode, *J. Physiol.* **145**, 320–370.

Fuortes, M. G. F. (1959). Initiation of impulses in visual cells of *Limulus, J Physiol.* **148**, 14–28.

Furshpan, E. J., and Potter, D. D. (1959). Slow postsynaptic potentials recorded from the giant motor fibre of the crayfish, *J. Physiol.* **115**, 326–335.

Furukawa, T., and Furshpan, E. J. (1963). Two inhibitory mechanisms in the Mauthner neurons of goldfish, *J. Neurophysiol.* **26**, 140–176.

Gerschenfeld, H. M., and Chiarandini, D. J. (1964). Ionic mechanism associated with non-cholinergic synaptic inhibition in molluscan neurons, *J. Neurophysiol.* **28**, 710–723.

Ginsborg, B. L. (1967). Ion movements in junctional transmission, *Pharmacol. Rev* **19**, 289–316.

Goldman, D. E. (1943). Potential, impedance, and rectification in membranes, *J. Gen. Physiol.* **27**, 37–60.

Gorman, A. L. F., and Marmor, M. F. (1970). Temperature dependence of the sodium-potassium permeability ratio of a molluscan neurone, *J. Physiol.* **210,** 919–931.

Hagiwara, S., Kusano, K., and Saito, S. (1960). Membrane changes in crayfish stretch receptor neuron during synaptic inhibition and under action of gamma-amino butyric acid, *J. Neurophysiol.* **23,** 505–515.

Hodgkin, A. L. (1948). The local electric changes associated with repetitive action in a non-medullated axon, *J. Physiol.* **107,** 165–181.

Hodgkin, A. L. (1951). The ionic basis of electrical activity in nerve and muscle, *Biol. Rev.* **26,** 339–409.

Hodgkin, A. L., and Horowicz, P. (1959). Movements of Na and K in single muscle fibres, *J. Physiol.* **145,** 405–432.

Hodgkin, A. L., and Huxley, A. F. (1939). Action potentials recorded from inside a nerve fibre, *Nature, Lond* **144,** 710–711.

Hodgkin, A. L., and Huxley, A. F. (1952*a*). Currents carried by sodium and potassium ions through the membrane of the giant axon of *Loligo, J Physiol.* **116,** 449–472.

Hodgkin, A. L., and Katz, B. (1949). The effect of sodium ions on the electrical activity of the giant axon of the squid, *J. Physiol.* **108,** 37–77.

Hodgkin, A. L., and Keynes, R. D. (1955). Active transport of cations in giant axons from *Sepia* and *Loligo, J Physiol.* **128,** 28–60.

Kerkut, G. A., and Thomas, R. C. (1964). The effect of anion injection and changes in the external potassium and chloride concentration on the reversal potentials of the IPSP and acetylcholine, *Comp. Biochem. Physiol.* **11,** 199–213.

Kimura, K., and Beidler, L. M. (1961). Microelectrode study of taste receptors of rat and hamster, *J. Cell. Comp. Physiol.* **58,** 131–140.

Koike, H., Brown, H. M., and Hagiwara, S. (1971). Hyperpolarization of a barnacle photoreceptor membrane following illumination, *J. Gen. Physiol.* **57,** 723–737.

Loewenstein, W. R., Terzuolo, C. A., and Washizu, Y. (1963). Separation of transducer and impulse-generating processes in sensory receptors, *Science* **142,** 1180–1181.

Nakajima, S., and Onadera, K. (1969). Membrane properties of the stretch receptor neurones of crayfish with particular reference to mechanisms of sensory adaptation, *J. Physiol.* **200,** 161–185.

Nastuk, W. L., and Hodgkin, A. L. (1950). The electrical activity of single muscle fibres, *J. Cell. Comp Physiol.* **35,** 39–73.

Oomura, Y., Ozaki, S., and Maéno, T. (1961). Electrical activity of a giant nerve cell under abnormal conditions, *Nature, Lond* **191,** 1265–1267.

Steinbach, H. B., and Spiegelman, S. (1943). The sodium and potassium balance in squid nerve axoplasm, *J. Cell. Comp. Physiol.* **22,** 187–196.

Tasaki, I., Watanabe, A., and Lerman, L. (1967). Role of divalent cations in excitation of squid giant axons, *Amer. J. Physiol.* **213,** 1465–1474.

Trautwein, W., and Dudel, J. (1958). Zum Mechanismus der Membranwirkung des Acetylcholin an der Herzmuskelfaser, *Pflüger's Arch ges. Physiol.* **266,** 324–334.

5

VOLTAGE CLAMPING

MEASUREMENT OF MEMBRANE CURRENTS

By 1949, Hodgkin and Katz had a pretty good idea how the nerve action potential worked. They were aware that the resting membrane was chiefly permeable to potassium and that an influx of sodium ions occurred during the action potential. The next step was to show *quantitatively* how the ion movements gave rise to the observed potentials. As mentioned in Chapter 1, the major obstacle to this theoretical treatment was the difficulty of measuring ionic currents directly. A large capacitive current, proportional to the rate of change of membrane potential, overlapped and masked the ionic part. To overcome this problem, Cole (1949) used a feedback circuit to hold the potential constant, eliminating the capacitive current.

This technique was subsequently taken up and developed by Hodgkin and Huxley, starting in 1952. The method is outlined schematically in Figure 1. The membrane potential is measured between an internal silver wire and an external lead. A command signal, E_c, is added to the external potential at the input of the control amplifier. The output of the amplifier is fed back to the inside of the axon via an attenuator. When connected in the manner shown, the feedback acts to reduce the difference in amplifier inputs to zero, or make $E_i - (E_o + E_c) = 0$. But this also makes $E_i - E_o = E_c$; that is, transmembrane potential is made equal to E_c. The amount of current necessary to do this is then measured across a series resistor from the bath to the circuit ground. Presumably, the injected current is equal and opposite to the current being generated by the active membrane. Usually a central portion of the nerve is insulated from the ends externally with

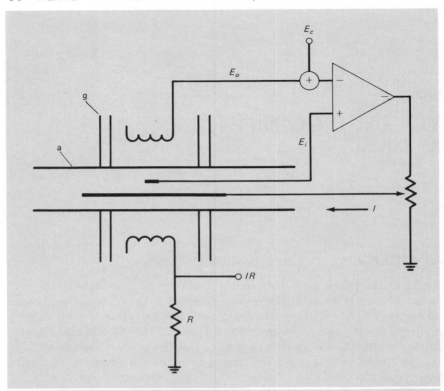

1

Voltage-clamp circuit for use with squid axon. a, axon; g, plastic guards. Uninsulated portion of internal voltage electrode indicated by short heavy line; uninsulated portion of current-injecting electrode indicated by long heavy line. E_i, internal potential; E_o, external potential; E_c, command potential. Feedback tends to make $E_i - E_o = E_c$. Injected current, I, measured as voltage drop across series resistor R.

plastic guards to limit the measured current to the area of membrane near the internal recording electrode; this method eliminates the problem of longitudinal spread of current. With regular pulses applied at E_c, the membrane potential starts out looking like the top trace in Figure 2, part A: an action potential rising from a passive charging curve. The measured current has the same shape as E_c, a square pulse. As the feedback is turned up (part B), the potential record is somewhat flattened, and irregularities appear in the current record. With sufficient feedback, the potential record resembles E_c, and the current has the form shown in part C. The first peak is required to charge the membrane capacitance quickly and produce the square corner at the start

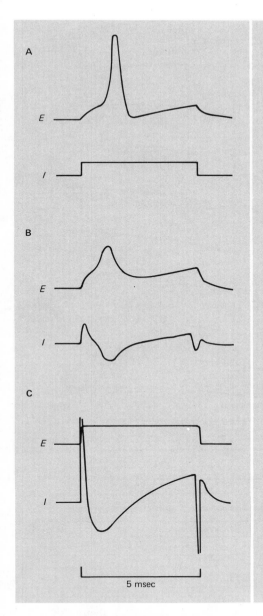

A

E

I

B

E

I

C

E

I

5 msec

Change from constant-current stimulation (A) through intermediate condition (B), to voltage-clamp condition (C), as feedback in Figure 1 is increased. *E,* membrane potential. *I,* injected current.

2

of the potential record. Next, an inward current is observed (shown in the downward direction), followed by an outward current. At the end of the command pulse, a downward transient is seen, as the membrane capacitance is suddenly discharged.

A series of currents measured in this manner in the squid axon by

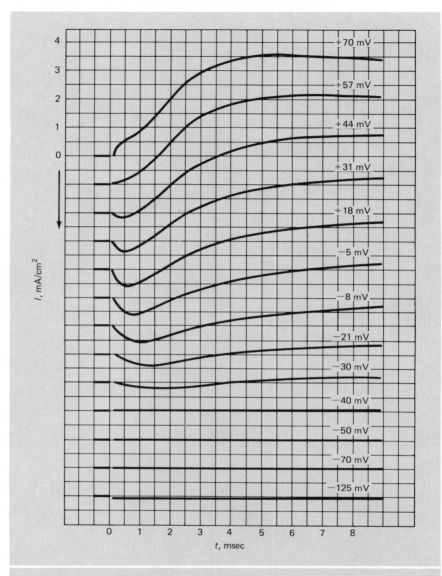

3

Currents measured with voltage clamp of squid axon. Inward currents indicated by downward deflections. Membrane held at about −60 mV (near resting potential), then stepped to potentials shown. (After Hodgkin *et al.*, 1952.)

Hodgkin *et al.* (1952) is shown in Figure 3. These traces show injected currents following steps of potential from the resting level to the voltages shown on the right of each line. (Only the response to application of a command pulse is shown, and not the response at the end of the command. The capacitive transients in Figure 2 have been removed, leaving gaps at time 0 in each trace.) At potentials more negative than −50 mV, little current is produced. Between −40 and about +40 mV, the direction of current flow is first inward through the membrane, then outward. Above +44 mV, the current is always outward.

By measuring the currents at fixed times after the start of the command pulse, it is possible to construct *I-V* curves, such as the ones shown in Figure 4. The currents at 0.5 and 8 msec are plotted as functions of the level to which the membrane potential is stepped by the command pulse. The *late* currents clearly illustrate the rectification mentioned in Chapter 4: at potentials more negative than −40 mV, the conductance, or value of $\Delta I/\Delta E$, is very low; this is referred to as the LEAK CONDUCTANCE, because it arises from leakage of ions through the resting membrane. At more positive potentials, the late currents, and hence the conductance, are greatly increased. This phenomenon is known as DELAYED RECTIFICATION.

The *I-V* curve for the early currents, however, is rather N-shaped; that is, it rises, falls, and then rises again. Between −40 and +20 mV, the current actually decreases as the membrane potential *increases*. This characteristic is known as NEGATIVE RESISTANCE, and it indicates a region of instability.

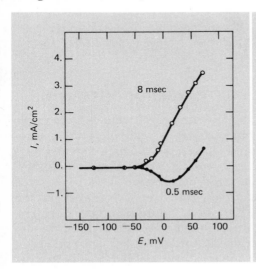

Current-voltage relation from voltage-clamp experiment in squid axon. (After Hodgkin *et al.*, 1952.)

4

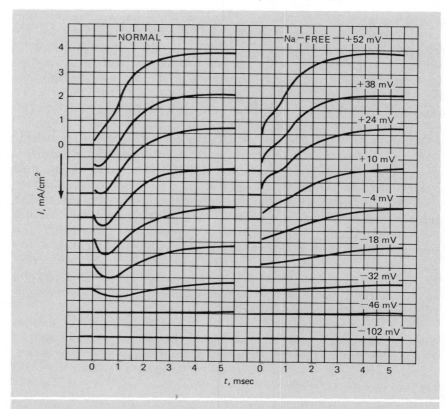

Currents measured in normal (left) and sodium-free solutions (right). Membrane potential held at −60 mV, then stepped to potentials shown on right. (After Hodgkin and Huxley, 1952*a*.)

SEPARATION OF SODIUM AND POTASSIUM CURRENTS

The above curves all show total membrane current during the nerve impulse. In order to separate out the sodium and potassium components of the current, Hodgkin and Huxley did the experiment whose data are shown in Figure 5 (based on data of Hodgkin and Huxley, 1952*a*). These traces are again currents (downward = inward) required to step the membrane potential from resting (about −60 mV) to the level shown. The records on the left were obtained in normal saline, and those on the right in a solution in which all the sodium was replaced with choline. This cation maintained the normal osmotic pressure of the external solution, but it was presumably not able to cross

5

the nerve membrane because of the large size of the molecule. The measured currents in the curves at potentials above $+10$ mV on the right in the figure are carried by potassium and outward-moving sodium. At less positive potentials there is little sodium outflow, and the outward-moving currents are equal to I_K. To obtain the variation of I_{Na} with time, the currents on the right are subtracted from those on the left at each level of depolarization. This is done in Figure 6 for the currents at -4 mV. The curve of $I_{Na} + I_K$ is taken from the left side of Figure 5. The vertical distance between these curves, equal to I_{Na}, is plotted below for comparison. This type of experiment enabled Hodgkin and Huxley to identify the entire *early inward current* with movement of sodium ions. Because the curves of $I_{Na} + I_K$ and I_K were superimposable at late times, the *late outward current* was identified with potassium ions.

To describe these results, Hodgkin and Huxley used sodium and potassium conductances defined as follows:

$$g_{Na} = \frac{I_{Na}}{E - E_{Na}} \tag{1}$$

$$g_K = \frac{I_K}{E - E_K} \tag{2}$$

where

$$E_{Na} = \frac{RT}{F} \ln \frac{[Na]_o}{[Na]_i} \approx +60 \text{ mV} \tag{3}$$

$$E_K = \frac{RT}{F} \ln \frac{[K]_o}{[K]_i} \approx -70 \text{ mV} \tag{4}$$

The variation of g_{Na} and g_K with time following a step depolariza-

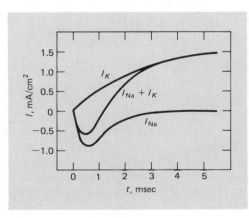

Voltage-clamp currents in squid axon measured in normal ($I_{Na} + I_K$) and sodium-free (I_K) solutions. I_{Na} calculated as difference between above two curves. (After Hodgkin and Huxley, 1952a.)

6

tion was similar to that of I_{Na} and I_K; that is, the sodium channel first became conductive, and then the potassium channel became conductive. The currents were considered to arise by flow of ions down their respective electrochemical gradients.

INSTANTANEOUS CURRENT-VOLTAGE RELATION

The above analysis suggests that the membrane follows Ohm's law and that a change in membrane current will be proportional to a change in potential. Of course, this test must be applied quickly after

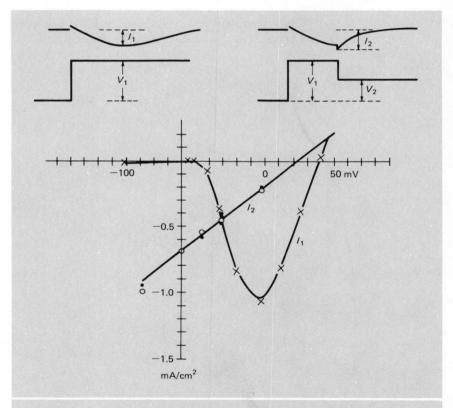

7

Instantaneous current-voltage relation obtained with voltage clamp for the early inward channel. X's indicate normal peak inward currents for various depolarizations. Solid dots and circles indicate variation of I_2 with V_2 as shown in inset on right. $I_2 - I_1$, instantaneous step of current produced by voltage step $V_1 - V_2$. Duration of first pulse = 1.53 msec. (After Hodgkin and Huxley, 1952b.)

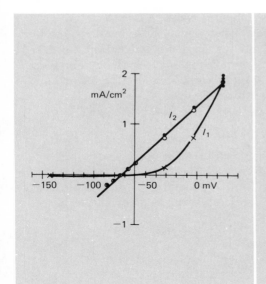

Instantaneous current-voltage relation for the late outward channel. Same method as in Figure 7, but duration of first depolarization long enough for full development of outward current. X's show normal currents produced by first step of potential. Solid dots and circles show I_2 versus V_2 as measured in Figure 7. $I_2 - I_1$ gives instantaneous step of current produced by voltage step $V_1 - V_2$. (After Hodgkin and Huxley, 1952b.)

changing the potential because we know the conductances start to vary a few tenths of a millisecond after a depolarization. The type of measurement used by Hodgkin and Huxley (1952b) to confirm the assumption of linearity (i.e., conformity to Ohm's law) is called the INSTANTANEOUS CURRENT-VOLTAGE RELATION and is measured as shown in Figure 7. These data apply to the early inward (sodium) channel particularly. The first pulse of depolarization (V_1) activates the sodium channel and produces an inward current I_1. Now, with the sodium conductance turned on the potential is stepped to V_2. The purpose of this is to measure the *change* in current $I_2 - I_1$ as a function of the *change* in potential $V_2 - V_1$ while the sodium conductance is activated. The X's on the graph show I_1, the inward current at various values of V_1. The values of I_2 are plotted versus V_2, following a V_1 of 29 mV. However, the I_2 line reflects the change $I_2 - I_1$, because I_1 is the same for all values of I_2. It can be seen that I_2 varies linearly with V_2, and so the instantaneous *I-V* relation is in fact linear. Also of interest is the reversal of the "tail" of current, I_2, at a potential near E_{Na} (about +40 mV). This indicates that the "tail" currents are carried by sodium.

In Figure 8 the same experiment is repeated for long first pulses, which last until after the early inward current is over and activate mainly the late outward (potassium) channel. Once again, the result was a linear *I-V* relation for second pulses, confirming Hodgkin and Huxley's formulation of the potassium conductance. In this case, however, the "tail" currents reversed near E_K (about -65 mV).

The striking symmetry and consistency of these results should not convince the reader that all nerves have linear instantaneous I-V curves; single nodes of Ranvier do not, for instance (Frankenhaeuser, 1960, 1963). In such cases a more complicated description of the sodium and potassium channels must be used.

ACTIVATION AND INACTIVATION

As seen in Figure 6, the sodium current following a step depolarization has an initial peak and then drops off after a few milliseconds, unlike the potassium current which builds up along a monotonic curve. Hodgkin and Huxley (1952c) considered that depolarization had two

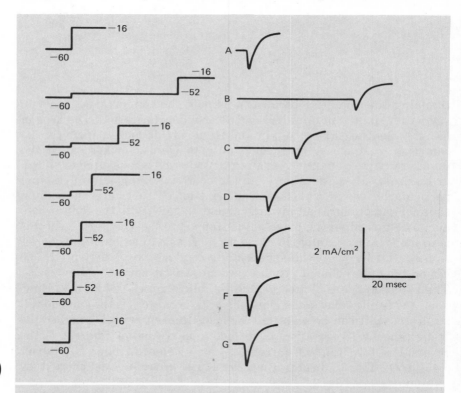

9

Time course of sodium inactivation. Left-hand column: Potentials under voltage-clamp conditions versus time. Right-hand column: Current versus time. (After Hodgkin and Huxley, 1952c.)

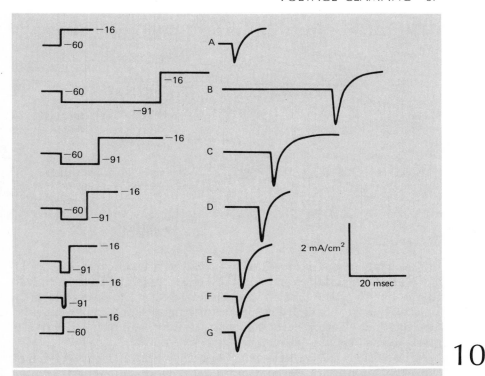

10

Time course of removal of resting sodium inactivation. Right and left columns show same measurements as in Figure 9. (After Hodgkin and Huxley, 1952c.)

effects on the sodium channel: ACTIVATION, or increased conductance, and INACTIVATION, which developed more slowly than the activation and "turned off" the channel. To find the time course of inactivation, they carried out experiments such as that shown in Figure 9. This is a two-step voltage-clamp experiment in which the first *conditioning* depolarization is always 8 mV, and the second *test* pulse always brings the membrane potential to 44 mV more depolarized than resting. As the duration of the first pulse is increased, the inward current in response to the second pulse begins to decrease. This inactivation develops with a half-time of a few milliseconds. If the first pulse is a hyperpolarization of 31 mV, the effect is a removal of resting inactivation, as shown in Figure 10. In this case, the inward current in response to the second step of potential (to 44 mV more depolarized than resting) is *augmented* above the one with no preceding pulse. Plots

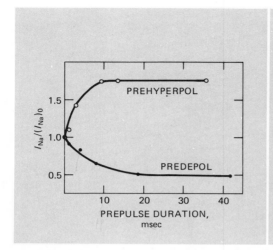

11

Time course of sodium inactivation. Curve marked PREDEPOL shows development of inactivation with conditioning depolarization of varying lengths. Curve marked PREHYPERPOL shows removal of resting inactivation with conditioning hyperpolarizations. Currents measured from data in Figures 9 and 10. (After Hodgkin and Huxley, 1952c.)

of development and removal of inactivation are shown in Figure 11. Both phenomena follow exponential curves, reaching maximal levels as the length of the first pulse is increased. The time constant of the inactivation in this figure is somewhat longer than that of removal of inactivation. These results demonstrate that sodium inactivation in the squid axon is *time-dependent.*

In addition, this process is *voltage-dependent* and varies strongly with the potential of the first pulse, as shown in Figure 12. This experiment had rather long (36-msec) conditioning pulses; so the effects developed maximally before the test pulses. With sufficiently large predepolarizations, as in A, the subsequent inward current in response to the test pulse was completely blocked; with no prepulse, as in E, a normal inward current was observed. Application of prehyperpolarization (F to I) augmented the inward current in response to the test pulse. To analyze this type of experiment, Hodgkin and Huxley made graphs such as that shown in Figure 13. The ratio of the inward current with each test pulse to that with no prepulse is plotted against the potential of the prepulse. Below about −90 mV, prepulses do not have much further effect on the test inward current. From −90 to −40 mV or so, the test current falls off sharply with increasingly positive prepulses. With those above −40 mV, the inward current is essentially blocked.

To describe this potential-dependent inactivation, Hodgkin and Huxley used an equation of the form

$$h = \frac{1}{1 + \exp\left[(V - V_h)/7\right]} \qquad (5)$$

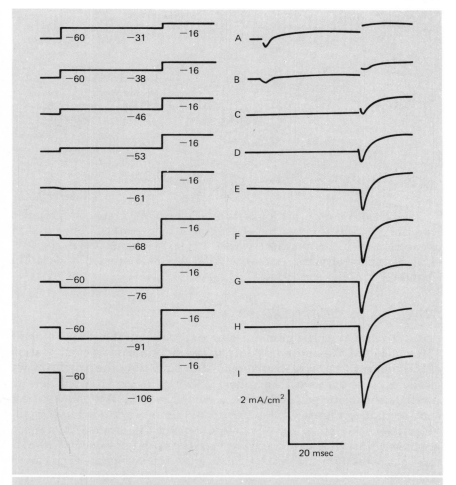

Voltage dependence of sodium inactivation. Prepulse of about 36 msec duration varied from 46-mV hyperpolarization to 29-mV depolarization, before a net test depolarization of 44 mV. (After Hodgkin and Huxley, 1952c.)

12

where

h = a parameter which is proportional to the inward sodium current

V = depolarization of prepulse measured from resting potential

V_h = value of V at which $h = \frac{1}{2}$

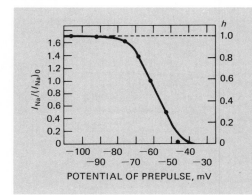

Voltage dependence of sodium inactivation. Normalized inward current in response to test pulse plotted against potential of conditioning pulse. Data from Figure 12. (After Hodgkin and Huxley, 1952c.)

The smooth curve in Figure 13 is a plot of h versus the prepulse potential. This is the first paper in which the quantity h appeared in connection with sodium inactivation. It plays a major role in the subsequent theory of nerve excitation, and will be discussed in detail in Chapter 6.

INTERNALLY PERFUSED AXONS

About 9 years after Hodgkin and Huxley's work burst forth in volumes 116 and 117 of the *Journal of Physiology*, an interesting confirmation of their ionic theory was discovered: After squeezing the axoplasm out of squid giant axons with a roller, Baker *et al.* (1961) were able to reinflate the axons with solutions of their choosing. Action potentials such as that in Figure 14, top part, could then be recorded in the perfused axons for up to several hours. Among the observations made possible with this technique were that (1) in the presence of 10-mM external potassium the resting potential of the perfused axons varied about 58 mV per tenfold change in *internal* [K] except at the lowest internal concentrations, and (2) the action potential was blocked by increasing the internal sodium concentration to the same as the external concentration.

Several laboratories then began to carry out voltage-clamp experiments using perfused axons (Moore *et al.,* 1963; Adelman and Gilbert, 1964; Adelman *et al.,* 1965; Armstrong and Binstock, 1965; and Tasaki and Singer, 1966). These and other studies with perfused axons have given many new insights into the process of nerve excitation, a topic which will be discussed in subsequent chapters. One might say that the most striking result of the perfusion studies has been the extent to which they confirm the sodium-potassium movements originally postulated by Hodgkin and Huxley.

EFFECTS OF CALCIUM ON MEMBRANE CURRENTS

The action of calcium on nerve membranes was first studied under voltage-clamp conditions by Frankenhaeuser and Hodgkin (1957), using the squid axon. They observed that reducing external calcium concentration caused a large increase in the inward current produced by a given step change of potential. When they plotted peak sodium conductance, defined as in Equation 1, versus potential to which the membrane was stepped, they obtained the curves shown in Figure 15. Reducing the external calcium by a factor of 5 had the effect of shifting the conductance-voltage curve about 15 mV toward more negative potentials. That is, in low-calcium solution the displacement of potential from resting which was necessary to obtain a given amount of inward current was about 15 mV less than in normal solution. A similar result was obtained for the late potassium current, as shown in Figure 16. In this case, reducing the external calcium concentration fivefold shifted the conductance-voltage curve for late current 10 to 15 mV more negative.

The authors summarized their results by saying that reducing the external calcium concentration fivefold had about the same effect as a depolarization of 10 to 15 mV; i.e., smaller changes in potential were then required to activate both inward and outward currents.

Frankenhaeuser and Hodgkin suggested that calcium might act by

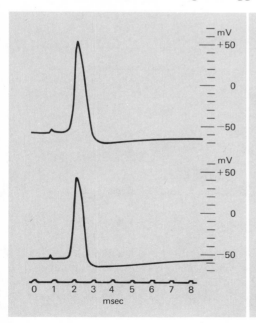

Action potentials in squid axon perfused with K_2SO_4 (top) and intact axon (bottom). (Baker *et al.*, 1961.)

14

binding to negative sites or carriers in the membrane and exerting a stabilizing influence. Outward (depolarizing) currents would "wash" the calcium ions off the sites by means of the imposed electric field and thus allow sodium to enter. The greater the external concentration of calcium, the more difficult it would be to remove calcium from the membrane, and the greater would be the threshold current, or depolarization, required to excite. This suggestion fits with experimental observations, but the exact mechanism is difficult to prove or disprove.

Blaustein and Goldman (1966) also measured shifts of the Na conductance-voltage curve along the voltage axis against changes in external calcium concentration. Hille (1968) published a curve of the variation of this shift with external calcium concentration in the node of Ranvier. Gilbert and Ehrenstein (1969), McLaughlin *et al.* (1971), D'Arrigo (1973), Brismar (1973), and others have suggested that calcium acts by forming a diffuse double layer outside the nerve mem-

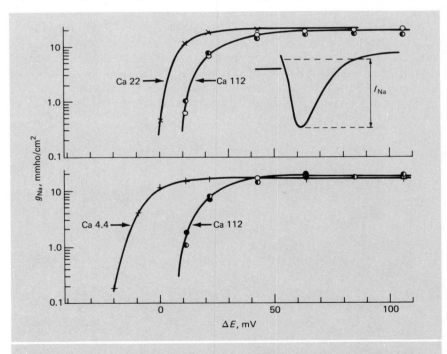

15

Effect of external calcium concentration on the variation of peak sodium conductance with membrane potential. (Calcium values next to curves slow external concentrations.) Potential held at resting level, then displaced by the amount shown on the abscissa. (After Frankenhaeuser and Hodgkin, 1957.)

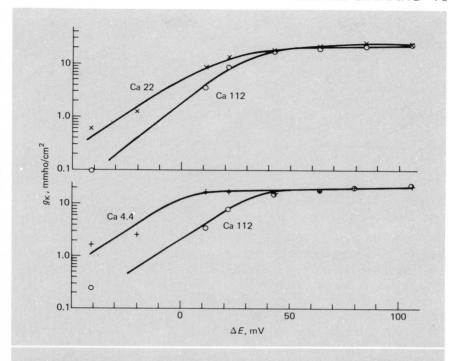

Effect of external calcium concentration on the variation of steady-state potassium conductance with membrane potential. Same method as in Figure 15. (After Frankenhaeuser and Hodgkin, 1957.)

16

brane, which can affect the transmembrane electric field. This idea is summarized in Figure 17. The lower line shows the normal distribution of potential across a resting membrane. The potential does not change abruptly at the inner or outer surface but is smoothed out by microscopic changes in ion concentrations between the membrane and the bulk solutions. Within the membrane, the rate of change of potential with distance, or FIELD STRENGTH, is assumed constant. The CRITICAL GRADIENT, or field strength required to produce an action potential, is shown by the dotted line. Normally depolarization, or reduction of E_m, lowers the gradient to this level and results in excitation. When the external calcium concentration is increased above normal, a layer of Ca^{++} ions forms near the membrane, loosely attracted by negative charges on the surface. This has the effect shown by the upper line: it makes the potential close to the outside surface more positive than normal, sharpening the corner of the potential curve. This change in the potential distribution can *increase* the transmembrane field

Schema to explain action of calcium on nerve membrane. Lower curve: Normal distribution of potential across membrane. Upper curve: Potential distribution with elevated external calcium concentration. $(dE/dx)_c$: Critical gradient for excitation. E_m: Measured transmembrane potential.

17

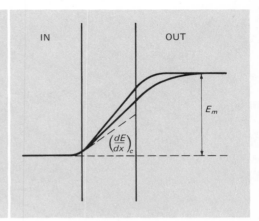

strength *without affecting* E_m. Thus, with elevated external calcium a greater-than-normal depolarization is required to reduce the field in the membrane to the critical level. While somewhat speculative, this type of model is useful in combining many of the observations of Ca^{++} action on nerve membranes, such as the lack of effect on resting potential and strong action on threshold.

One of the next big steps in our understanding of the role of calcium will, of course, be to discover its intracellular functions in perfused axons.

PROBLEMS

1. Using the data of Figure 3, plot membrane current at 0.5 msec after the start of the command pulse as ordinate (outward current is upward) versus potential to which the membrane is stepped (depolarizations to the right). On the same graph, plot the currents measured at 8 msec after the start of the pulse. At what potential does the early inward current pass through zero and become outward?

2. Using the normal data in Figure 5 and the same current and voltage conventions as in Problem 1, plot the currents at 0.5 and 5.5 msec after the start of the stimulus versus membrane potential. On the same graph plot the corresponding currents in Na-free solution. Does replacement of external sodium affect the late outward current?

3. For a depolarization to -4 mV, measure and tabulate the currents in Figure 5 in normal and Na-free solutions versus time, using 0.25-msec intervals. From these data, plot I_{Na} and I_K versus time on the same graph.

4. In Problem 3, E_{Na} was $+60$ mV and E_K was -70 mV. Calculate g_{Na} and g_K (in mmho/cm²) versus time, and plot. What is the peak value of g_{Na}? The final value of g_K?

REFERENCES

Adelman, W. J., Dyro, F. M., and Senft, J. (1965). Long duration responses obtained from internally perfused axons, *J. Gen. Physiol.* **48**, 1-9.

Adelman, W. J., and Gilbert, D. L. (1964). Internally perfused squid axons studied under voltage clamp conditions, *J. Cell. Comp. Physiol.* **64**, 423-428.

Armstrong, C. M., and Binstock, L. (1965). Anomalous rectification in the squid giant axon injected with tetraethylammonium chloride, *J. Gen. Physiol.* **48**, 859-872.

Baker, P. F., Hodgkin, A. L., and Shaw, T. I. (1961). Replacement of the protoplasm of a giant nerve fibre with artificial solutions, *Nature, Lond.* **190**, 885-887.

Blaustein, M. P., and Goldman, D. E. (1966). Competitive action of calcium and procaine on lobster axon, *J. Gen. Physiol.* **49**, 1043-1063.

Brismar, T. (1973). Effects of ionic concentration on permeability properties of nodal membrane in myelinated nerve fibres of *Xenopus laevis;* potential clamp experiments, *Acta Physiol. Scand.* **87**, 474-484.

Cole, K. S. (1949). Dynamic electrical characteristics of the squid axon membrane, *Arch. Sci. Physiol.* **3**, 253-258.

D'Arrigo, J. S. (1973). Possible screening of surface charges on crayfish axons by polyvalent metal ions, *J. Physiol.* **231**, 117-128.

Frankenhaeuser, B. (1960). Quantitative description of sodium currents in myelinated nerve fibres of *Xenopus laevis, J. Physiol.* **151**, 491-501.

Frankenhaeuser, B. (1963). A quantitative description of potassium currents in myelinated nerve fibres of *Xenopus laevis, J. Physiol.* **169**, 424-430.

Frankenhaeuser, B., and Hodgkin, A. L. (1957). The action of calcium on the electrical properties of squid axons, *J. Physiol.* **137**, 218-244.

Gilbert, D. L., and Ehrenstein, G. (1969). Effect of divalent cations on potassium conductance of squid axons: determination of surface charge, *Biophysic. J.* **9**, 447-463.

Hille, B. (1968). Charges and potentials at the nerve surface. Divalent ions and pH, *J. Gen. Physiol.* **51**, 221-236.

Hodgkin, A. L., and Huxley, A. F. (1952a). Currents carried by sodium and potassium ions through the membrane of the giant axon of *Loligo, J. Physiol.* **116**, 449-472.

Hodgkin, A. L., and Huxley, A. F. (1952b). The components of membrane conductance in the giant axon of *Loligo, J. Physiol.* **116**, 473-496.

Hodgkin, A. L., and Huxley, A. F. (1952c). The dual effect of membrane potential on sodium conductance in the giant axon of *Loligo, J. Physiol.* **116**, 497-506.

Hodgkin, A. L., Huxley, A. F., and Katz, B. (1952). Measurement of current-voltage relations in the membrane of the giant axon of *Loligo, J. Physiol.* **116**, 424-448.

Hodgkin, A. L., and Katz, B. (1949). The effect of sodium ions on the electrical activity of the giant axon of the squid, *J. Physiol.* **108**, 37-77.

McLaughlin, S. G. A., Szabo, G., and Eisenman, G. (1971). Divalent ions and the surface potential of charged phospholipid membranes, *J. Gen. Physiol.* **58**, 667-687.

Moore, J. W., Narahashi, T., and Ulbricht, W. (1963). Sodium conductance shift in an axon internally perfused with a low K solution, *Fed. Proc.* **22,** 174.

Tasaki, I., and Singer, I. (1966). Membrane macromolecules and nerve excitability: a physico-chemical interpretation of excitation in squid giant axons, *Ann. N.Y. Acad. Sci.* **137,** 792–806.

6

THE HODGKIN-HUXLEY MODEL
AND OTHER THEORIES
OF EXCITATION

PREDICTION OF VOLTAGE-CLAMP CURRENTS

It is important to note that there are really two parts to the HODGKIN-HUXLEY theory: (1) Their parallel-conductance model of the axon membrane, which, in conjunction with the voltage-clamp experiments, yielded so much valuable information, and (2) the mathematical model which was constructed to describe the voltage and time dependencies of the conductances (Hodgkin and Huxley, 1952*d*). Other, more recent models also attempt to account for the known properties of the Na and K currents, which are well established. The original Hodgkin-Huxley theory was intended to account for the following observations from the voltage-clamp experiments: (1) separate currents carried by Na$^+$ and K$^+$ ions, (2) a potassium conductance which turns on with an S-shaped time course following a step depolarization, and (3) a sodium conductance which turns on and then inactivates (Hodgkin and Huxley, 1952*d*).

The starting assumption for this theory is that there are separate channels for sodium, potassium, and other ions and that the transmembrane current in the voltage clamp is given by

$$I = C_m \frac{dE}{dt} + g_{Na}(E - E_{Na}) + g_K(E - E_K) + g_l(E - E_l) \tag{1}$$

where

$$
\begin{aligned}
I &= \text{current density, A/cm}^2 \\
E &= \text{membrane potential, V} \\
C_m &= \text{membrane capacity, F/cm}^2 \\
g_{Na} &= \text{sodium conductance, mho/cm}^2 \\
g_K &= \text{potassium conductance, mho/cm}^2
\end{aligned}
$$

g_l = leak conductance, mho/cm^2
E_{Na} = sodium equilibrium potential, V
E_K = potassium equilibrium potential, V
E_l = leak equilibrium potential, V

This equation describes the current across a nervous structure whose interior is isopotential (no longitudinal spread of current). The conductances g_{Na} and g_K are assumed to vary with potential and time in a deterministic way. The leak conductance is assumed constant. Under voltage-clamp conditions, the rate of change of potential is zero, and the membrane currents are purely ionic and should be given by the last three terms of Equation 1. In order to see if the model can predict real membrane currents, it is first necessary to write expressions for g_{Na} and g_K as functions of potential and time.

THEORETICAL POTASSIUM AND SODIUM CONDUCTANCES

The potassium conductance is described by

$$g_K = \bar{g}_K n^4 \qquad (2)$$

where

$\bar{g}_K$ = maximum potassium conductance, a constant
n = a dimensionless variable which varies from 0 to 1 as a function of voltage and time

The fourth power of n is needed to describe the slow buildup of g_K following a step depolarization. The physical analogue of such a process is that four particles must be near a certain area of the membrane at once in order for a potassium ion to cross, where the probability of each particle being there is proportional to n. Hodgkin and Huxley worked out an equation for n as a function of potential and time, which need not be derived here. This allowed them to calculate g_K versus the time following a step depolarization.

The sodium channel is a little more complicated because it inactivates. In order to include this in their theory, Hodgkin and Huxley assumed that

$$g_{Na} = \bar{g}_{Na} m^3 h \qquad (3)$$

where

$\bar{g}_{Na}$ = maximum sodium conductance, a constant
m = an activation parameter like n, which varies from 0 to 1 as a function of voltage and time
h = an inactivation parameter which varies from 0 to 1 as a function of voltage and time

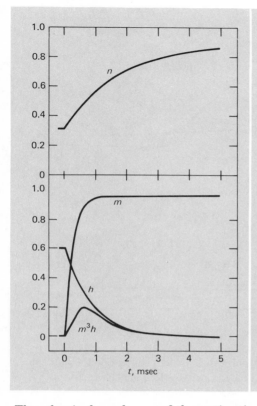

Variation of Hodgkin-Huxley parameters with time following a step depolarization in a voltage-clamped axon.

1

The physical analogue of the activation and inactivation processes is that three m particles and one h particle must be near an area of membrane for a sodium ion to cross.

The variation of n, m, and h following a step depolarization in a voltage-clamped axon is shown in Figure 1. n builds up exponentially with time, giving rise to the observed potassium conductance. m builds up exponentially with a short time constant, and h falls off exponentially with a long time constant. The product m^3h then has the shape indicated, which corresponds to the time course of the sodium conductance. When the theoretical values of n, m, and h are plugged into Equation 1, the predicted currents are quite close to those measured with the voltage clamp.

In addition, the theory could predict the shape of *action potentials* in the length-clamped axon (in which the inside is made to be isopotential by means of a metal electrode). To do this, Hodgkin and Huxley solved Equation 1 by numerical methods under the condition that E was allowed to vary. Three such solutions are shown in the top part of Figure 2. The "theoretical" axon was first given short shocks,

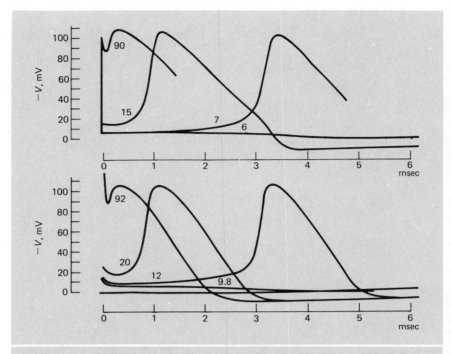

2

Calculated and observed action potentials in length-clamped axon. Top part: Theoretical action potentials following short shocks to the axon. Bottom part: Measured action potentials following similar stimulation of real axon. (Hodgkin and Huxley, 1952*d*.)

whose intensities are indicated by the numbers by each trace. Then the membrane potential was calculated as a function of time. The smallest stimulus produced a slight depolarization, which then decayed back to the resting potential. A slightly larger stimulus finally gave rise to a spike, which reached a height of over 100 mV from resting. Stronger depolarizations caused the spike to occur earlier, but the overshoot was relatively constant. The bottom of Figure 2 shows some action potentials from real axons, for comparison with the model results. The predictions are fairly close quantitatively, as well as in form. [It should be noted that these curves were obtained for an axon at 6°C, much colder than the normal ocean temperature at which the squid live (perhaps 12 to 18°C). Hodgkin and Huxley used low temperatures to slow down the axon responses, as an aid to obtaining a good voltage clamp. The theoretical values of n, m, and h would have to have much shorter time constants at higher ambient temperatures.]

PREDICTION OF PROPAGATED ACTION POTENTIAL

The above theoretical conductances were calculated for length-clamped axons which responded with the same transmembrane potential in all active areas. However, Hodgkin and Huxley were able to apply their theory to axons which were *not* length-clamped, where the membrane potential could vary with distance along the axon. To do this, they used the fact that the same equations governing the behavior of a whole axon under length clamp also apply to a small patch of axon membrane. Thus, the current through a patch is given by

$$I = C_m \frac{dE}{dt} + \overline{g}_K n^4 (E - E_K) + \overline{g}_{Na} m^3 h (E - E_{Na}) + \overline{g}_l (E - E_l) \quad (4)$$

where

$I =$ current density, A/cm², a function of x and t
    (x, distance from a fixed point on the axon)
$E =$ transmembrane potential, V, a function of x and t
$\overline{g}_K$, $\overline{g}_{Na}$, and $\overline{g}_l =$ peak conductances, constants, mho/cm²
n, m, and h are functions of E and t

From the cable theory (see Hodgkin and Rushton, 1946), the current through a patch of membrane is given by

$$I = \frac{a}{2R_a} \frac{d^2E}{dx^2} \quad (5)$$

where

$a =$ radius of the axon, cm
$R_a =$ specific resistivity of the axoplasm, ohm-cm

From Equations 4 and 5, it is possible to write a partial differential equation for E at any point on the axon. However, E is expressed in terms of t in Equation 4 and in terms of x in Equation 5, which makes the solution very difficult. A trick may now be used to circumvent this problem: Because the action potential is known to be propagated with a constant velocity, it may be expressed as

$$E = f(x\text{-}\theta t) \quad (6)$$

where $\theta =$ conduction velocity, cm/sec

This is an expression of a traveling wave, and it has the property that if the potential at point x has a certain time course, the potential at a point θt away from x will have the *same* time course as that at x, but delayed by t. Now, if we let $w = x - \theta t$, we can find the *spatial* derivative $\delta^2 E/\delta x^2$ in terms of the *time* derivative $\delta^2 E/\delta t^2$ (Slater and Frank, 1947):

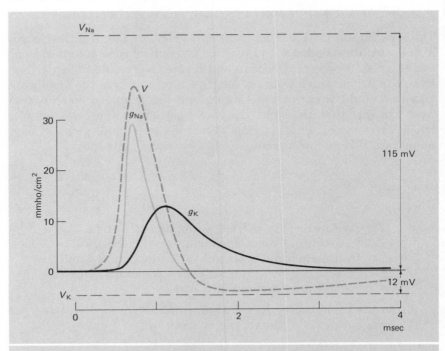

Theoretical changes in sodium and potassium conductances and membrane potential during a propagated nerve impulse. (Hodgkin, 1964.)

3

$$\frac{\delta E}{\delta x} = \frac{\delta E}{\delta w} \tag{7}$$

$$\frac{\delta^2 E}{\delta x^2} = \frac{\delta^2 E}{\delta w^2} \tag{8}$$

$$\frac{\delta E}{\delta t} = -\theta \frac{\delta E}{\delta w} \tag{9}$$

$$\frac{\delta^2 E}{\delta t^2} = \theta^2 \frac{\delta^2 E}{\delta w^2} \tag{10}$$

From Equations 8 and 10,

$$\frac{\delta^2 E}{\delta x^2} = \frac{1}{\theta^2} \frac{\delta^2 E}{\delta t^2} \tag{11}$$

Equation 4 can now be written as a function of time only, using Equations 5 and 11:

$$\frac{a}{2R_a\theta^2}\frac{d^2E}{dt^2} = C_m\frac{dE}{dt} + \bar{g}_K n^4(E-E_K) + \bar{g}_{Na}m^3h(E-E_{Na}) + \bar{g}_l(E-E_l) \qquad (12)$$

The spatial variable x has now been removed from the potential equation. This is possible because all solutions for any x are the *same function of time,* just delayed by a certain amount.

Equation 12 can now be solved by assuming values of the conduction velocity, θ, and solving by an iterative process. For most values of θ, the potential becomes infinite after an action potential. The correct value of θ is that for which the potential returns to resting after an impulse.

The behaviors of the sodium and potassium conductances can also be studied during the action potential, and are shown in Figure 3. V_{Na} and V_K are the sodium and potassium equilibrium potentials. g_{Na} and g_K are the theoretical sodium and potassium conductances. V is the action potential calculated from Equation 12, where $V = E - E_r$ (E_r, resting potential). The propagated action potential is not substantially different from that in the length-clamped axon; both have about the same amplitude, and both exhibit an UNDERSHOOT, or hyperpolarization following the spike, to a level slightly more negative than resting.

THRESHOLD AND OTHER MEMBRANE PHENOMENA

The Hodgkin-Huxley equations could thus account for the time course of the propagated action potential in terms of the known variation of membrane conductances with potential and time. These equations could also explain some other membrane phenomena, including the threshold, or critical depolarization, for production of the action potential. Because the sodium conductance was increased transiently by depolarization and because increased g_{Na} in turn *leads* to further depolarization, the change in membrane potential during a spike is *regenerative.* This may be schematized as

The slower-rising potassium conductance has the opposite effect, namely to hyperpolarize the membrane.

The THRESHOLD potential may be defined as that potential at which the inward Na current is just equal to the outward K current. If the K current is larger, the potential will return to resting. If the Na current is larger, the potential will become unstable and produce a spike.

Furthermore, the Hodgkin-Huxley mathematical model could account for REFRACTORINESS, or the decreased excitability of an axon following an action potential. Two factors contributed to this property: (1) the fact that g_{Na} was somewhat inactivated following a spike and so could not reach as large a value in response to a second stimulus and (2) the persistence of the potassium conductance following a spike (see Figure 3). The amount of g_K which remains activated gives rise to the after-hyperpolarization, or undershoot, of the action potential and also makes it more difficult to achieve a sodium current which exceeds the potassium current. The time course of the refractory period in the squid axon is quite similar to those of g_K and h, the Na inactivation parameter (Hodgkin and Huxley, 1952d). Thus, the refractory period may be understood in terms of known potential and time variations of the conductances.

Other membrane phenomena which the Hodgkin-Huxley equations can explain include ANODE-BREAK EXCITATION and ACCOMMODATION. "Anode-break" refers to removal of a hyperpolarizing stimulus, which, in the squid axon and other nerve membranes, often gives rise to an action potential. This is predicted from the removal of resting Na inactivation by the hyperpolarizing pulse (see Chapter 5). Accommodation is most often used to mean an increase in threshold with decreasing rate of rise of a stimulus (see, for example, Araki and Otani, 1959; and Bradley and Somjen, 1961). Occasionally it is used to mean slowing of repetitive discharge during a maintained stimulus. In either case, it is simply explained by Na inactivation and K activation during a depolarizing stimulus.

The success of the Hodgkin-Huxley equations in predicting these and other membrane phenomena showed the utility of mathematical modeling of this type. Even though the time and voltage dependencies of the conductances were known from experiment, it was not obvious that they would work together in concert to predict the action potential, threshold, refractoriness, and accommodation. The model showed that they could do so.

CONFIRMATION OF PREDICTED ION MOVEMENTS BY TRACER STUDIES

Further verification of the ionic theory was provided by measurements of actual influxes, using radioisotopes. In 1951, Keynes showed that the extra net sodium influx during stimulation of cephalopod axons was about 4×10^{-12} mol/cm² per impulse. Calculation of the net Na influx from the Hodgkin-Huxley equations gave approximately the same result.

An even more convincing demonstration of the sodium influx during activity was provided by Atwater *et al.* (1969), who measured it *during* a voltage-clamp experiment on an internally perfused squid axon. They were able to show that the measured sodium influx per impulse was at least 0.92 times the computed ionic flux and that the potential at which the measured ionic flux disappeared was the same as the reversal potential for the early inward current. All of these and more recent experiments make it difficult not to accept the idea that an action potential consists of an early inward current which is carried by sodium and a later outward current which is carried by potassium. Yet this notion was not accepted and could not be verified until the development of the squid-axon preparation and voltage-clamp technique.

MODIFICATION OF THE MODEL FOR THE NODE OF RANVIER

The behavior of single Ranvier nodes under voltage-clamp conditions is best described by a slightly different formulation than that given by Hodgkin and Huxley for the squid axon. Frankenhaeuser (1960, 1963) found that the instantaneous *I-V* relations for sodium and potassium for single Ranvier nodes were not linear (as in Figures 7 and 8, Chapter 5). This meant that Equations 1 and 2, Chapter 5, could not be used to describe the Na and K currents. Instead, Frankenhaeuser used expressions for the currents derived from the constant-field theory (Hodgkin and Katz, 1949):

$$I_{Na} = P_{Na} \frac{F^2 E}{RT} \frac{[Na]_o - [Na]_i \exp (EF/RT)}{1 - \exp (EF/RT)} \tag{13}$$

$$I_{K} = P_{K} \frac{F^2 E}{RT} \frac{[K]_o - [K]_i \exp (EF/RT)}{1 - \exp (EF/RT)} \tag{14}$$

where

P = permeability
F = Faraday constant
E = membrane potential
R = universal gas constant
T = temperature, °K

To describe the kinetics of the sodium and potassium currents, Frankenhaeuser used the forms

$$P_{Na} = \overline{P}_{Na}m^2h \tag{15}$$

$$P_K = \overline{P}_K n^2 k \tag{16}$$

where

m and h have the same meanings as in the Hodgkin-Huxley model
n is a potassium activation parameter, as in the Hodgkin-Huxley model
k is a slow inactivation parameter for the potassium current

Other than the m^2 dependence for the sodium channel and the potassium inactivation factor, the behavior of the single node was well described by this version of the Hodgkin-Huxley theory.

ALTERNATIVE MODELS

While the mathematical model of Hodgkin and Huxley predicts the ionic currents in the squid axon quite well, it is not the only model which can do so. Another approach was that of Hoyt (1963), who used empirical relationships for g_{Na} and g_K which were chosen to fit the observed currents exactly. For instance, g_K was assumed to be a function of v_K, a parameter like the Hodgkin-Huxley n, having the form

$$v_K = v_{K_\infty}(1 - e^{-\alpha_K t}) \tag{17}$$

where

v_{K_∞} = final value of v_K
α_K = inverse time constant

Then v_K was chosen to give the correct values of g_K at different potentials and times. It turned out that for small values of g_K the empirical relationship was

$$g_K = \overline{g}_K(v_K + C)^p \tag{18}$$

where

$\overline{g}_K$ = maximum K conductance
C = constant
p = constant exponent

This is to be contrasted with the Hodgkin-Huxley relationship, where g_K was proportional to n^4, as in Equation 2. For larger values of g_K, the curve was less rapidly rising with v_K than in Equation 18. For the sodium channel, instead of assuming that g_{Na} depended on two parameters, such as m and h, she assumed that it varied with one parameter, v_{Na}, which had the form

$$v_{Na} = A(1 - e^{-at}) - B(1 - e^{-bt}) \tag{19}$$

v_{Na} was then chosen empirically to give the correct values of g_{Na} at different potentials and times.

This model can predict every behavior of the squid axon which is predicted by the Hodgkin-Huxley equations. It gives *better* fits to the g_{Na} and g_K curves, because it is constructed empirically *from* those conductances. In addition, Hoyt's model can account for the very slow rise of K current following a strong hyperpolarization (see Chapter 7), which is not a feature of the original Hodgkin-Huxley theory.

Goldman (1964) took a very different approach from that of Hodgkin and Huxley; he assumed that the ion carriers in the membrane are the *same* for sodium as for potassium. The carriers are thought to have binding sites which can *exchange* Na^+ for K^+ and also Ca^{++}. This is outlined as shown:

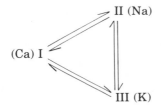

The sites may be in any of three states: In state I, the resting condition, the sites preferentially bind calcium. Depolarization causes many of the sites to pass into state II, and bind sodium. This is the state which gives rise to the early inward current in the voltage clamp. In state III the sites bind potassium, and this produces the late outward current. Finally, the sites return to state I. It is assumed that there are n_{Na} sites per square centimeter in the Na state, n_K in the K state, and n_{Ca} in the Ca state, at any time. Goldman calculated the potential and time dependence of the n's, and from this the variations of n_{Na} and

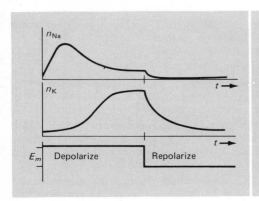

Variation of number of sodium-binding (n_{Na}) and potassium-binding (n_K) sites per square centimeter with time following a depolarization. Follows Goldman's theory of excitation (1964). E_m: membrane potential.

4

n_K following a step depolarization. These are shown in Figure 4. While no quantitative comparison was made to Hodgkin and Huxley's experimental data, the similarity to the curves of g_{Na} and g_K in the voltage-clamped squid axon is striking.

In both of these alternative models, different approaches from that of Hodgkin and Huxley were used to predict the *same* voltage-clamp currents. These currents are a permanent fact about squid axon membranes and will always have to be explained by new theories. Perhaps because of priority, the Hodgkin-Huxley analysis has remained the most popular and has been applied to giant axons in marine annelids (Goldman and Schauf, 1973) and lobsters (Julian *et al.*, 1962*b*); to neuron cell bodies (Connor and Stevens, 1971*a, b, c*); to muscles in vertebrates (Adrian *et al.*, 1970) and invertebrates (Hagiwara *et al.*, 1968); and to heart muscle (Morad and Orkand, 1971), to name a few. It is not exaggerating to say that the Hodgkin-Huxley theory has dominated the study of excitation for almost 20 years. This eminence was well-deserved, for the theory was constructed with care and insight. Now and for the next several years, as modern molecular approaches are used to reveal the machinery of excitation, the observations of Hodgkin and Huxley will continue to be as good as the best available, and their theory will be regarded as a valuable guide.

PROBLEMS

1. After a particular step depolarization in Hodgkin and Huxley's model axon, the parameter n follows the curve

$$n = 0.891 - 0.576e^{-t/1.7}$$

   where t is in milliseconds. The value of $\bar{g}_K$ in the model is 24.3 mmho/cm². Plot g_K as a function of time for 10 msec, using steps of 0.5 msec for the fast-rising part (up to 4 msec). What is the final value of g_K?

2. After the same depolarization as in Problem 1, the parameters m and h in the Hodgkin-Huxley model follow the curves

$$m = 0.963(1 - e^{-t/0.252})$$
$$h = 0.605e^{-t/0.84}$$

   The value of $\bar{g}_{Na}$ is 70.7 mmho/cm². Plot g_{Na} as a function of time for 10 msec, using steps of 0.25 msec up to 2 msec. What is the largest value of g_{Na} reached?

3. From the Hodgkin-Huxley explanation of nerve accommodation, what would you predict about the amplitude (overshoot) of an action potential produced by a slowly rising ramp stimulus, compared with that produced by a rapidly rising stimulus? (For experimental results, see Vallbo, 1964.)

4. Calculate the amount of ions in mol/cm² which must be transferred across a membrane with a capacitance of 1 μF/cm² to change the potential 100

mV, or about the same amount as an action potential. The charge across a capacitor is given by $q = CV$, where $q =$ charge separation in coulombs, $C =$ capacitance in farads, and $V =$ potential in volts. The amount of coulombs per mole of ions is 96,500, which can be assumed to be 10^5. Can you account for any discrepancy between the answer to this problem and the amount of charge found by Keynes for net sodium entry during an action potential?

5. After a step depolarization in the Hoyt model of the squid axon, the sodium parameter, v_{Na}, follows the curve

$$v_{Na} = 71.0(1 - e^{-t/0.219}) - 53.7(1 - e^{-t/1.20})$$

g_{Na} varies monotonically with v_{Na}. Plot the time course of v_{Na} for the first 10 msec after the start of the depolarization. What is the largest value of v_{Na} reached?

REFERENCES

Adrian, R. H., Chandler, W. K., and Hodgkin, A. L. (1970). Voltage clamp experiments in striated muscle fibres, *J. Physiol.* **208**, 607–644.

Araki, T., and Otani, T. (1959). Accommodation and local response in motoneurones of toad's spinal cord, *Japan. J. Physiol.* **9**, 69–83.

Atwater, I., Bezanilla, F., and Rojas, E. (1969). Sodium influxes in internally perfused squid giant axon during voltage clamp, *J. Physiol.* **201**, 657–664.

Bradley, K., and Somjen, G. G. (1961). Accommodation in motoneurones of the rat and the cat, *J. Physiol.* **156**, 75–92.

Connor, J. A., and Stevens, C. F. (1971a). Inward and delayed outward membrane currents in isolated neural somata under voltage clamp, *J. Physiol.* **213**, 1–19.

Connor, J. A., and Stevens, C. F. (1971b). Voltage clamp studies of a transient outward membrane current in gastropod neural somata, *J. Physiol.* **213**, 21–30.

Connor, J. A., and Stevens, C. F. (1971c). Prediction of repetitive firing bebehavior from voltage clamp data on an isolated neurone soma, *J. Physiol.* **213**, 31–53.

Frankenhaeuser, B. (1960). Quantitative description of sodium currents in myelinated nerve fibres of *Xenopus laevis, J. Physiol.* **151**, 491–501.

Frankenhaeuser, B. (1963). A quantitative description of potassium currents in myelinated nerve fibres of *Xenopus laevis, J. Physiol.* **169**, 424–430.

Goldman, D. E. (1964). A molecular structural basis for the excitation properties of axons, *Biophysic. J.* **4**, 167–188.

Goldman, L., and Schauf, C. L. (1973). Quantiative description of sodium and potassium currents and computed action potentials in *Myxicola* giant axons, *J. Gen. Physiol.* **61**, 361–384.

Hagiwara, S., Takahashi, K., and Junge, D. (1968). Excitation-contraction coupling in a barnacle muscle fiber as examined with voltage clamp technique, *J. Gen. Physiol.* **51**, 157–175.

Hodgkin, A. L. (1964). *The Conduction of the Nervous Impulse*, Springfield, Ill., Thomas, p. 63.

Hodgkin, A. L., and Huxley, A. F. (1952*d*). A quantitative description of membrane current and its application to conduction and excitation in nerve, *J. Physiol.* **117**, 500–544.

Hodgkin, A. L., and Katz, B. (1949). The effect of sodium ions on the electrical activity of the giant axon of the squid, *J. Physiol.* **108**, 37–77.

Hodgkin, A. L., and Rushton, W. A. H. (1946). The electrical constants of a crustacean nerve fibre, *Proc. Roy, Soc. B.* **133**, 444–479.

Hoyt, R. C. (1963). The squid giant axon. Mathematical models, *Biophysic. J.* **3**, 399–431.

Julian, F. J., Moore, J. W., and Goldman, D. E. (1962*b*). Current-voltage relations in the lobster giant axon membrane under voltage clamp conditions, *J. Gen. Physiol.* **45**, 1217–1238.

Keynes, R. D. (1951). The ionic movements during nervous activity, *J. Physiol.* **114**, 119–150.

Morad, M., and Orkand, R. K. (1971). Excitation-contraction coupling in frog ventricle: evidence from voltage clamp studies. *J. Physiol.* **219**, 167–189.

Slater, J. C., and Frank, N. H. (1947). *Mechanics,* New York, McGraw-Hill, p. 163.

Vallbo, A. B. (1964). Accommodation related to inactivation of the sodium permeability in single myelinated nerve fibres from *Xenopus laevis, Acta Physiol. Scand.* **61**, 429–444.

7

INDEPENDENCE OF THE SODIUM
AND POTASSIUM CHANNELS

The Hodgkin-Huxley formulation contains no dependency, or COU-PLING, between the sodium and potassium channels — they are treated as separate and distinct mechanisms. (For example, in Equation 2, Chapter 5, I_K does not depend on E_{Na}, or vice-versa.) Several lines of evidence appear to support the idea of separate channels.

INDEPENDENT EFFECTS OF POTENTIAL ON SODIUM AND POTASSIUM CURRENTS

It is possible to alter the magnitude and time course of both the sodium and potassium currents in voltage-clamped axons by conditioning depolarization or hyperpolarization. The effect of predepolarization is to inactivate the inward current without affecting the outward current (see Figures 9 to 13, Chapter 5). Cole and Moore (1960) showed that large conditioning hyperpolarizations (up to -212 mV absolute membrane potential) delay the onset of the outward current with little or no effect on the inward current. These outward currents, obtained from a two-step voltage-clamp experiment, are shown in Figure 1. First, the membrane was hyperpolarized for 3 msec to a potential of -52 to -212 mV; then it was depolarized to $+60$ mV (near the sodium equilibrium potential). The larger the prehyperpolarization, the greater was the delay of the potassium currents. In order to account for this delay, Cole and Moore found that it was necessary to use an expression of the form

$$I_K = \overline{g}_K n^{25}(E - E_K) \tag{1}$$

instead of the fourth-power dependency used by Hodgkin and Huxley (cf. Equation 2, Chapter 6).

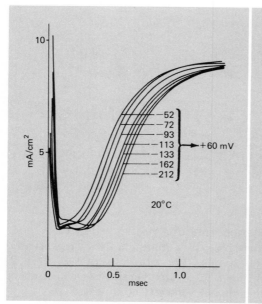

1

Effect of prehyperpolarization on outward currents in
voltage-clamped squid axon.
(Cole and Moore, 1960.)

The independent variability of the inward and outward currents thus suggests that they are carried by separate mechanisms.

DIFFERENT ION SELECTIVITIES OF EARLY AND LATE CHANNELS

Different ion selectivities may be inferred by changing external ion concentrations around active neurons under voltage clamp and looking at the effects on early and late currents. As shown by Moore *et al.* (1966), the early inward-current channel in the squid axon admits sodium and lithium but excludes potassium and rubidium. The late outward-current channel admits potassium and rubidium but excludes sodium and lithium. The same relationships are seen in the node of Ranvier (Hille, 1972). Hence, it appears there are two separate mechanisms for the early and late currents, with completely different selectivities.

PHARMACOLOGICAL SPECIFICITIES OF THE EARLY AND LATE CHANNELS

Strong support for the idea of separate channels came from the fortuitous discoveries of agents which specifically block one channel with-

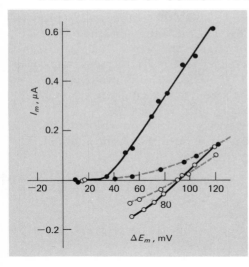

Effect of tetraethylammonium ion on outward currents in *Onchidium* neuron. I_m: voltage clamp currents, E_m: membrane potential. Circles show early currents, solid dots late currents. Solid lines obtained in normal saline, gray lines in saline containing about 90-mM TEA. (Hagiwara and Saito, 1959.)

2

out affecting the other. In 1959, Hagiwara and Saito examined the effects of tetraethylammonium (TEA) on a voltage-clamped neuron. TEA has the following structure:

$$C_2H_5-N^+-C_2H_5$$

with C_2H_5 above and C_2H_5 below the nitrogen.

Some of the results are shown in Figure 2. The circles show the early inward current and the solid dots the late outward current. Application of TEA (gray lines) greatly reduced the outward current but had

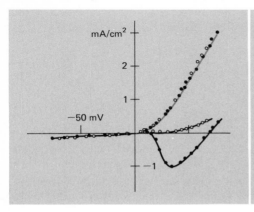

Effect of tetrodotoxin on inward currents in squid axon. Solid lines show early currents, gray line late currents. Solid dots were obtained in normal saline, circles in saline containing 5 × 10^{-7} g/ml TTX. (Nakamura *et al.*, 1965.)

3

little effect on the inward current. This differential action has also been demonstrated in the squid axon (Armstrong and Binstock, 1965). A more recent analysis has shown that in the node of Ranvier the only Hodgkin-Huxley parameter affected by TEA is $\bar{g}_K$, the maximum potassium conductance (Hille, 1967). The effect of this compound on unclamped neurons is to prolong the action potential greatly by delaying and inhibiting the active repolarization.

Tetrodotoxin (TTX) is a neuroactive poison from the puffer fish; it has an analogous action on the early inward-current channel. The TTX molecule has the following structure:

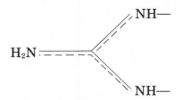

In 1965 Nakamura *et al.* carried out the first voltage-clamp study of TTX, using the squid axon. One result is shown in Figure 3. The solid line with solid dots shows the normal early inward current, and the solid line with circles is the early current (now outward) after application of TTX. The gray line is the late outward current, which is unaffected by TTX.

Tetrodotoxin has a highly specific blocking action on the sodium channel in almost all excitable membranes where it has been tried. This may be due to interaction of the guanidinium group

$$H_2N \begin{array}{c} \diagup NH— \\ \diagdown NH— \end{array}$$

with the membrane Na channel. The TTX molecule may bind to the Na-selective carrier but is too large to pass through the membrane. The selective actions of TTX and TEA on nerve and muscle membranes thus imply that the compounds are acting on separate Na- and K-carrying systems.

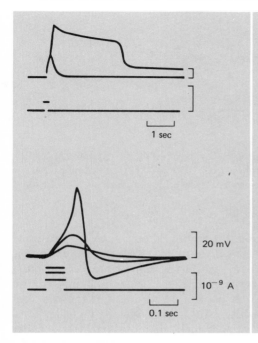

Differential development of spike plateau and repolarization mechanisms in tunicate muscle. Upper traces in each part show membrane potential, lower traces applied current. Top part: Response of muscle cell of tadpole with 90% grown tail. Bottom part: Response of muscle cell of fully grown tadpole after hatching. Same voltage and current calibrations in all parts. (Takahashi *et al.*, 1971.)

4

DIFFERENTIAL DEVELOPMENT OF INWARD AND OUTWARD CHANNELS

Although no voltage-clamp work has been done on the preparation, Takahashi *et al.* (1971) have presented evidence on differential development of spike depolarization and repolarization processes in tunicate muscle. As shown in Figure 4, top part, the first all-or-none action potential which appears in the muscle has a plateau of several seconds' duration. At a later stage of development, shown in the bottom part, the active repolarization mechanism has become strong enough to shorten the spike to less than 0.1 sec. The temporal difference in development of the depolarizing and repolarizing processes suggests that they may involve separate mechanisms.

MULLINS' CRITERION AND THE ELIMINATION OF THE SODIUM INACTIVATION PROCESS

In 1968, Mullins stated a possible test for the existence of separate channels for the early and late currents in nerve: Hodgkin and Huxley (1952d) measured the peak sodium conductance, $\bar{g}_{Na}$, as 120 mmho/cm² and the peak potassium conductance, $\bar{g}_{K}$, as 36 mmho/cm². (These

5

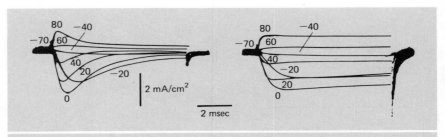

Destruction of Na inactivation by perfusion of squid axon with pronase. Left side: Voltage-clamp currents with 15-mM TEA in internal perfusate. Right side: Same conditions after a 6-min perfusion with pronase. (Armstrong *et al.,* 1973.)

values are also reported as about 100 mmho/cm² *each* by Mullins.) However, the point was made that if there are two separate channels, then it should be possible to open both at the same time and measure a total membrane conductance greater than 120 mmho/cm², for instance. The fact that no one had been able to do so suggested that there was only one channel, which could carry Na, K, or other ions.

In 1973, this criterion was apparently met by Armstrong *et al.,* using the perfused squid axon. They found that in axons perfused for the correct length of time with the proteolytic enzyme pronase the sodium conductance did not inactivate. This is shown in Figure 5. The left side shows voltage-clamp currents obtained in the presence of TEA, which are presumably carried by sodium. On the right side, after treatment of the axon with the enzyme the sodium currents clearly do not inactivate. If the axon was treated with pronase in the *absence* of TEA, then the late total membrane current was larger than normal because of maintained activation of the sodium current. Thus, when the sodium current stays turned on, it can add to the potassium current. This finding is at odds with the single-channel assumption, which says that potassium channels are a subsequent form of sodium channels and are created by inactivation of sodium channels.

SINGLE-CHANNEL MODELS

The above studies support the concept of separate channels very strongly. However, the case should not be considered entirely closed, because at least three distinguished workers (Mullins, 1959, 1968; Goldman, 1964, 1969; and Tasaki, 1968) have taken issue with the dual-channel idea. They prefer to consider that basically the same molecular machinery carries both sodium and potassium currents.

Mullins' model regards the channels as pores in the membrane whose size can change, causing a variation in selective permeability from potassium to sodium and back again. Goldman's concept is that phospholipid dipoles in the membrane rotate during a depolarization and that this changes the affinity of the membrane sites from calcium to sodium and then to potassium. Tasaki considers that the membrane contains a calcium-sodium-potassium ion exchanger, rather than separate mechanisms for carrying different ionic currents. In particular, he points out that the parallel-conductance model used by Hodgkin and Huxley is electrically indistinguishable from a *single* voltage source in series with a *single* conductance if the source and conductance are assumed to have certain voltage and time dependencies.

ARGUMENTS FOR A SINGLE CHANNEL

All these coupled models have the feature that the sodium conductance in a single channel must turn off before the potassium conductance can turn on, because both ionic species are carried in the same channel. This assumption seems to conflict with the independent variability of the sodium and potassium currents which can be obtained with prehyperpolarization or depolarization, as mentioned above. However, Carnay and Tasaki (1971) point out that not all the channels must "flip" from one state to the next at the same time; so variable overlap of the Na and K currents is possible with a single-channel model. This is illustrated in Figure 6. The curves on top are the time courses of g_{Na} and g_K in the squid axon, according to Hodgkin and Huxley. The patterned squares underneath are hypothetical "snapshots" of a mosaic membrane consisting of resting sites (gray dots), Na binding sites (circles), and K binding sites (solid dots) at different times. In A, the resting state of the membrane, most of the sites are in the low-conductance, resting state. In B, during the rising phase of the action potential, most of the sites have switched to the Na-selective state, although a few have gone over to the K-selective state. In C, during the active repolarization phase, most of the sites have gone from the Na-selective to the K-selective state. In D, the sites return to the low-conductance, resting state. As indicated by the conductance-time curves in Figure 6, this mosaic model allows for temporal separation of Na and K currents using the same channel, or carrier, for both Na^+ and K^+ ions. In fact, nothing about this model prohibits the reduction of g_{Na} without affecting g_K, or the great delaying of the onset of g_K, as mentioned earlier in this chapter.

The apparent specificity of action of TTX and TEA can also apply to a one-channel model if the channel is assumed to bind a particular

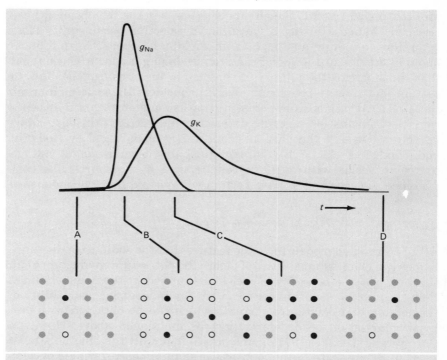

6

Explanation of temporal separation of Na and K currents in terms of single-channel model. Top part: Time course of g_{Na} and g_K. Bottom part: mosaic representation of membrane. ◉: resting state of channel, ○: N_a-selective state, ●: K-selective state. A−D shows membrane at different phases of action potential. (After Carnay and Tasaki, 1971.)

inhibitor when in one state and not when in a different state. That is, the channel might bind TTX when in the Na-selective state and re-lease the TTX in favor of potassium when it passes to the K-selective state. A similar argument might apply to other differences between the channels. The different selectivities, for instance, could serve as a description of a single channel which shows different properties at different times.

Thus, it seems difficult to claim with certainty that two completely separate mechanisms exist for the movement of sodium and potassium ions during the nerve impulse in a normal axon. The above observa-tions are merely intended as a *caveat* to the reader to maintain an open mind as further developments come in.

PROBLEMS

1. Using the values of n from Problem 1, Chapter 6, and $\bar{g}_K = 274$ mmho/cm^2, plot the quantity

$$g_K = \bar{g}_K n^{25}$$

for a period of 10 msec, using 1-msec steps up to 4 msec. What is the final value of g_K?

2. Moore *et al.* (1967) showed that in order to block the sodium channels in some lobster nerves, it takes about 1.6×10^{-11} mol of tetrodotoxin (TTX) per gram of nerve. How many TTX *molecules* per gram of nerve are required to block?

3. Light- and electron-microscope studies have shown that these lobster nerves have a total membrane area of about 0.7×10^4 cm^2/g. How many TTX molecules per square micron of membrane are required to block the sodium channels? (If *at least* one TTX molecule is required to block each sodium channel, this figure represents an upper limit to the number of Na channels per square micron of membrane.)

4. As mentioned by Tasaki (1968), the squid axon membrane may be represented by a *single* channel as shown:

ART WORK

What are the values of g_m and E_m in terms of the g_{Na}, g_K, g_{Cl}, E_{Na}, E_K, and E_{Cl} in Figure 4, Chapter 4?

REFERENCES

Armstrong, C. M., Bezanilla, F., and Rojas, E. (1973). Destruction of sodium conductance inactivation in squid axons perfused with pronase, *J. Gen. Physiol.* **62**, 375–391.

Armstrong, C. M., and Binstock, L. (1965). Anomalous rectification in the squid giant axon injected with tetraethylammonium chloride, *J. Gen. Physiol.* **48**, 859–872.

Carnay, L. D., and Tasaki, I. (1971). Ion exchange properties and excitability of the squid giant axon, *Biophysics and Physiology of Excitable Membranes,* W. J. Adelman (ed.), New York, Van Nostrand Reinhold, 379–422.

Cole, K. S., and Moore, J. W. (1960). Potassium ion current in the squid giant axon: dynamic characteristic, *Biophysic. J.* **1**, 1–14.

Goldman, D. E. (1964). A molecular structural basis for the excitation properties of axons, *Biophysic. J.* **4**, 167–188.

Goldman, D. E. (1969). Physico-chemical models of excitable membranes, *The Molecular Basis of Membrane Function,* D. C. Tosteson (ed.), Englewood Cliffs, N. J., Prentice-Hall, 259–279.

Hagiwara, S., and Saito, N. (1959). Voltage-current relations in nerve cell membrane of *Onchidium verruculatum, J. Physiol.* **148**, 161–179.

Hille, B. (1967). The selective inhibition of delayed potassium currents in nerve by tetraethylammonium ion, *J. Gen. Physiol.* **50,** 1287–1302.

Hille, B. (1972). The permeability of the sodium channel to metal cations in myelinated nerve, *J. Gen. Physiol.* **59,** 637–658.

Hodgkin, A. L., and Huxley, A. F. (1952*d*). A quantitative description of membrane current and its application to conduction and excitation in nerve, *J. Physiol.* **117,** 500–544.

Moore, J. W., Anderson, N., Blaustein, M., Takata, M., Lettvin, J. Y., Pickard, W. F., Bernstein, T., and Pooler, J. (1966). Alkali cation selectivity of squid axon membrane, *Ann. N.Y. Acad. Sci.* **137,** 818–829.

Moore, J. W., Narahashi, T., and Shaw, T. I. (1967). An upper limit to the number of sodium channels in nerve membrane? *J. Physiol.* **188,** 99–105.

Mullins, L. J. (1959). An analysis of conductance changes in squid axon, *J. Gen. Physiol.* **42,** 1013–1035.

Mullins, L. J. (1968). A single channel or a dual channel mechanism for nerve excitation, *J. Gen. Physiol.* **52,** 550–553.

Nakamura, Y., Nakajima, S., and Grundfest, H. (1965). The action of tetrodotoxin on electrogenic components of squid giant axons, *J. Gen. Physiol.* **48,** 985–996.

Takahashi, K., Miyazaki, S., and Kidokoro, Y. (1971). Development of excitability in embryonic muscle cell membranes in certain tunicates, *Science* **171,** 415–418.

Tasaki, I. (1968). *Nerve Excitation: A Macromolecular Approach,* Springfield, Ill., Thomas.

8

DIVALENT IONS
AS CHARGE CARRIERS

DIVALENT SPIKES AND BI-IONIC POTENTIALS

The sodium hypothesis of nerve and muscle excitation was so important and convincing that for a number of years scientists continued to find only sodium-dependent action potentials in excitable membranes. However, in 1953 Fatt and Katz observed that action potentials could be seen in crab muscle fibers in Na-free solutions, as long as calcium or magnesium ions were present in the external solution. They suggested that Ca^{++} or Mg^{++} might act as the charge carrier.

In 1958, Fatt and Ginsborg performed some ion substitutions using crayfish muscle fibers and concluded that neither sodium nor magnesium was essential for the production of action potentials; only external calcium was indispensable. These authors also found that replacement of Ca^{++} with Sr^{++} or Ba^{++} caused even larger action potentials than were seen with the normal amount of Ca^{++}. Consequently, they developed a theory of divalent action potentials based largely on strontium spikes. They assumed that when all the external calcium was replaced with strontium, the principal inward current through the membrane was due to Sr^{++} and the principal outward current to K^{+}. By equating these currents at the peak of the action potential, they obtained the relationship

$$\frac{P_{Sr}[Sr]_o}{P_K[K]_i} = \frac{\exp{(EF/RT)}[\exp{(EF/RT)} + 1]}{4} \tag{1}$$

where

$$P_{Sr} = \text{strontium permeability}$$
$$P_K = \text{potassium permeability}$$

$[Sr]_o$ = external strontium concentration
$[K]_i$ = internal potassium concentration
E = membrane potential at peak of spike
R = universal gas constant
T = temperature, °K
F = Faraday constant

While this equation cannot be solved explicitly for E, it permits plotting $[Sr]_o$ as a function of E. One feature of this equation is that the slope of the curve of E versus $\ln [Sr]_o$ changes with E. (This is not true of a "pure-strontium" electrode, for which E is given by

$$E = \frac{RT}{2F} \ln \frac{[Sr]_o}{[Sr]_i} \tag{2}$$

In this case the slope of E versus $\ln [Sr]_o$ is constant.) In Equation 1, when E is large and negative,

$$\frac{P_{Sr}[Sr]_o}{P_K[K]_i} = \frac{\exp (EF/RT)}{4} \tag{3}$$

and

$$E = \frac{RT}{F} \ln \frac{4P_{Sr}[Sr]_o}{P_K[K]_i} \tag{4}$$

This has a slope of 58 mV per tenfold change in $[Sr]_o$. When E is large and positive, Equation 1 becomes

$$\frac{P_{Sr}[Sr]_o}{P_K[K]_i} = \frac{\exp (2EF/RT)}{4} \tag{5}$$

and

$$E = \frac{RT}{2F} \ln \frac{4P_{Sr}[Sr]_o}{P_K[K]_i} \tag{6}$$

This has a slope of 29 mV per tenfold change in $[Sr]_o$, like Equation 2. The variation of E with $[Sr]_o$ over a large range is shown in Figure 1.

This treatment was one of the earliest applications of the BI-IONIC POTENTIAL THEORY (where inward current was carried by one ion and outward current by another *at the same instant*) to a biological membrane. In Hodgkin and Huxley's theory, both early inward and early outward currents were carried by sodium; the later outward potassium current did not much overlap the sodium currents in time. In Fatt and Ginsborg's theory, by contrast, the inward Sr current and outward K current overlapped considerably; so both affected the action-potential peak.

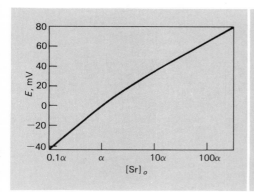

Graph of Equation 1, showing theoretical variation of action-potential overshoot, E, with external strontium concentration, $[Sr]_o$. $\alpha = P_K[K]_i/2P_{Sr}$, defined in Equation 1. (Fatt and Ginsborg, 1958.)

CALCIUM SPIKES

In 1964, Hagiwara, Naka, and others in California began to explore the ionic dependency of the action potential in the giant muscle fiber of the barnacle, *Balanus nubilus* Darwin. This exceedingly convenient preparation consisted of single muscle cells which often attained 4 cm in length and 2 mm in diameter. These huge fibers could easily be impaled with longitudinal electrodes, and the internal sarcoplasm could be replaced with buffered solutions having desired ion concentrations.

A few of the early observations which Hagiwara *et al.* made about the barnacle muscle fibers included (1) the fact that usually no all-or-none action potential was produced by depolarizing these fibers in normal conditions, but that (2) when the internal Ca concentration was reduced by injecting some calcium chelating agent such as EGTA [ethylene glycol bis (β-amino-ethylether)-N,N'-tetraacetic acid], then all-or-none spikes could be reliably produced. Also, (3) replacement of external sodium with tris (tris-hydroxymethyl aminomethane) had no effect on the all-or-none spike, and (4) the amplitude of the spike varied with *external* Ca concentration and *internal* K concentration in perfused fibers. This last behavior is illustrated in Figure 2. Columns A, B, and C were obtained from fibers injected with solutions of decreasing K concentration, 485, 48, and 0-mM, respectively. Rows, 1, 2, and 3 were obtained from these fibers in external solutions containing, respectively, 20, 85, and 338-mM Ca. The spikes were very much prolonged with K-free internal solutions, but in all cases the overshoot varied directly with log $[Ca]_o$ and inversely with log $[K]_i$. This result indicated that the membrane permeability to K^+ was substantial compared with that of Ca^{++} during the peak of the spike.

Some other properties of calcium spikes which were unlike those of sodium spikes had been observed by Hagiwara and the earlier work-

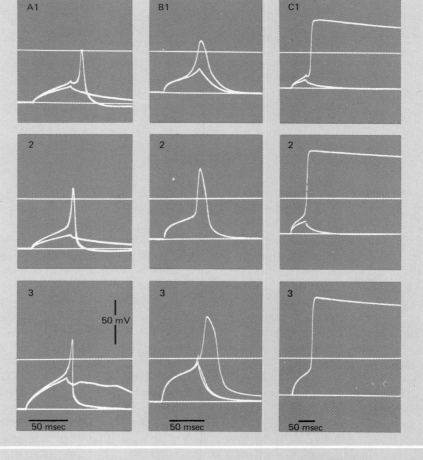

2

Variation of barnacle muscle action potential with external calcium concentration and internal potassium concentration. Subthreshold stimuli applied in several pictures. (Hagiwara *et al.*, 1964.)

ers, including that (1) tetrodotoxin (TTX) had no effect on the action potentials, even at concentrations up to 1 mM, (2) procaine increased the amplitude of the action potential, and (3) Co^{++}, Mn^{++}, Ni^{++}, and La^{+++} inhibited the action-potential mechanism at relatively low concentrations.

About the same time as the group in California was studying Ca spikes in barnacles, this type of action-potential mechanism also began to emerge in molluscan nerve cells (Gerasimov *et al.*, 1964; Meves, 1966; and Kerkut and Gardner, 1967). In each case with molluscan neurons, the spikes followed the criteria of (1) insensitivity to external Na concentration or tetrodotoxin, (2) variation with external [Ca], [Ba], or [Sr], and (3) blockage by external Co, Mn, Ni, or La. [Barium, in addition to substituting for calcium as an inward current carrier, has the property of greatly prolonging the action potential in many preparations. Sperelakis *et al.* (1967) have related this effect to a specific reduction of the potassium conductance by barium.]

The calcium spike was established as an entity by 1965 and has since been suggested for such structures as mussel heart (Irisawa *et al.*, 1967) and catch muscle (Twarog, 1967), tunicate muscle (Miyazaki *et al.*, 1972), and cell bodies of snail neurons (Wald, 1972).

MIXED Na–Ca SPIKES

Meanwhile, some examples were appearing prior to 1965 of nerves and muscles in which the spike amplitude depended on both external sodium and calcium — the so-called MIXED-DEPENDENCY class of spikes. Koketsu *et al.* (1959) found this behavior in frog spinal ganglion cells; Bülbring and Kuriyama (1963) in smooth muscle; and Niedergerke and Orkand (1966*a, b*) in heart muscle. Then Kerkut and Gardner (1967) and Junge (1967) described some molluscan neurons in which both sodium and calcium contributed to the action-potential amplitude. This is illustrated in Figure 3, obtained with the R2 cell in *Aplysia*. The top row shows the action-potential amplitude in normal saline (NS), sodium-free (Tris) solution, and normal saline afterward. The middle row shows the spike in normal saline, calcium-free (Tris) solution, and normal afterward. The bottom row shows normal, sodium- and calcium-free, and normal afterward. An all-or-none spike could be obtained in Na-free or Ca-free solutions for hours. However, replacement of both the external Na and Ca blocked the spike within 5 min, and the spike returned upon replacement of Na in the external solution. This mixed-dependency spike in *Aplysia* evidently arises from two separate ionic channels, based on three lines of evidence: (1) The Na spike in Ca-free solution is blocked by TTX but not by cobalt, (2) the Ca spike in Na-free medium is blocked by cobalt but not by TTX (Geduldig and Junge, 1968), and (3) Geduldig and Gruener (1970) observed inward Ca currents in voltage-clamped *Aplysia* neurons in Na-free solutions or in the presence of tetrodotoxin. The Ca currents were about one-tenth as large as the Na currents and had a significantly slower time course. A similar separation of Na and Ca

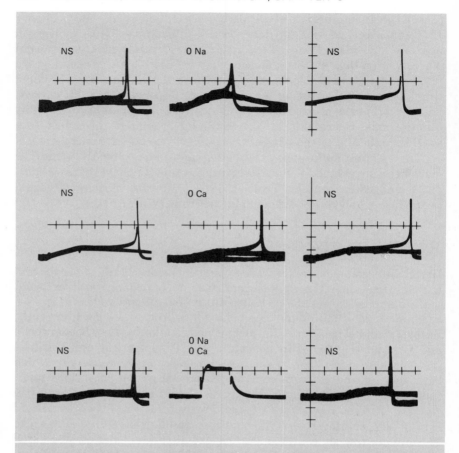

3

Variation of spike overshoot with external sodium and calcium concentration in *Aplysia* cell body. Sodium and calcium replaced with Tris. NS, normal saline; 0 Na, sodium-free; 0 Ca, calcium-free. Subthreshold stimuli applied in several pictures. (Geduldig and Junge, 1968.)

currents into functionally different channels was suggested for neurosecretory cells in the crayfish eyestalk x-organ (Iwasaki and Satow, 1971).

CALCIUM CURRENT IN SQUID SODIUM CHANNEL

The independent behavior of the Na and Ca channels in the above neurons is to be contrasted with that of the inward-current channel in the squid axon, which admits both Na and Ca. Watanabe *et al.* (1967)

demonstrated that action potentials could be produced in this preparation in Na-free solutions. The axons were first perfused with a proteolytic enzyme to remove as much axoplasm as possible, and were then
injected with a solution of 25 to 100-mM CsF, made isotonic with
glycerol. The results are shown in Figure 4. The top row of pictures
shows action potentials obtained when the external solutions were
300-mM hydrazinium chloride plus 200-mM CaCl$_2$ (A); 100-mM guanidinium chloride, 200-mM tetramethylammonium (TMA), and 200-mM
CaCl$_2$ (B); 10-mM KCl, 290-mM TMA, and 200-mM CaCl$_2$ (C); and
200-mM BaCl$_2$ made isotonic with glycerol (D). Action potentials
could be produced in all these solutions as long as some *divalent* ions
were present in the external medium. The bottom row shows the effect
of addition of 2×10^{-8} g/ml tetrodotoxin to the external solutions. All
the action potentials except that obtained with KCl outside were completely blocked. Watanabe *et al.* felt that TTX was blocking the usual
inward-current channel and that therefore all the ions contributing to
the action potentials, including Ca^{++}, must be passing through that
channel.

This curious result has been confirmed and explained convincingly
in a recent voltage-clamp study by Meves and Vogel (1973). These
authors show *two* components of inward Na current in the squid axon:

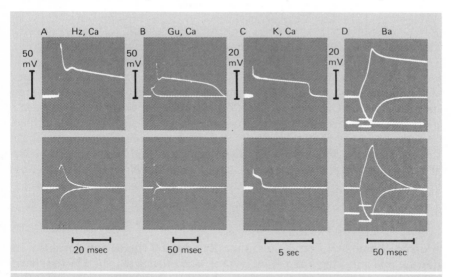

4

Spike activity in squid axon in sodium-free solutions. Top row: Various
solutions as described in text. Bottom row: Effect of tetrodotoxin on action potentials in above solutions. (Watanabe *et al.*, 1967.)

5

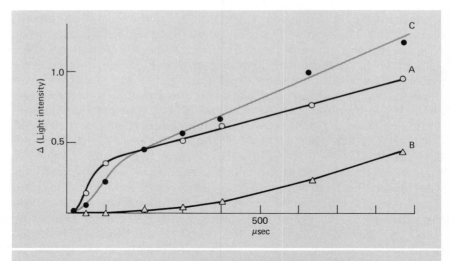

Two phases of calcium entry in squid axon. Aequorin injected into axon emits light with increasing internal Ca^{++} concentration. Curve A: Light in response to voltage-clamp commands of increasing duration. Curve B: Same after application of 0.8-μM TTX. Curve C: After removal of TTX. Δ(light intensity) is light normalized with respect to that produced by action potential, to compensate for changes with time. (Baker *et al.*, 1971.)

The first corresponds to Hodgkin and Huxley's inward current (up to a few milliamperes per square centimeter), and the second is much longer-lasting (time constant 14 msec at 17°C) and much smaller (tens of microamperes per square centimeter). In Na-free solutions containing 100-mM CaCl$_2$, Meves and Vogel observed inward *Ca currents* with the same time course as the slow phase of the Na current. The Ca currents were blocked by TTX and were presumably going through the Na channel. These slow divalent currents could explain the types of action potentials seen in the squid axon in Na-free solutions by Watanabe *et al.* (1967).

AEQUORIN

A direct demonstration of calcium entry during the action potential in nerves and muscles has recently been obtained with the Ca-sensitive protein, aequorin. This substance, obtained from a luminescent jellyfish, is injected into excitable cells and emits light in the presence of ionized calcium. Thus, by recording the emitted light with a photomultiplier tube it is possible to follow dynamic changes in intracellular

Ca concentration. Entry of Ca^{++} ions during activity has been shown by this method in the squid axon by Baker *et al.* (1971). Two components of light emission were seen during stimulation of an axon which had been injected with aequorin; this is shown in Figure 5. Curve A is the light resulting from a step depolarization as a function of the duration of the step. This shows the development of the rapid phase within about 200 μsec. Curve B was obtained after application of TTX, which eliminates the rapid component but leaves intact a slow phase of Ca entry. Baker *et al.* showed that the rapid phase had the same time course as the inward Na current. This, together with the TTX result, indicated that the rapid phase was due to Ca permeability of the Na channel (about 1% of the Na permeability). The slow phase was shown to resemble the known Ca channel in presynaptic terminals of the squid stellate synapse (Katz and Miledi, 1967, 1969).

Baker *et al.* (1971) point out that both the *synaptic channel* and the *slow axonal channel* are insensitive to tetrodotoxin and tetraethylammonium (TEA) and both are blocked by Mg^{++} or Mn^{++} ions in the external solution. This slow axonal channel is evidently *not* the same as Meves and Vogel's (1973) divalent channel discussed above, because Meves and Vogel's Ca currents were blocked by tetrodotoxin.

Aequorin injection has also been used to demonstrate calcium entry in the barnacle muscle fiber (Ashley and Ridgway, 1970), in presynaptic terminals in the squid stellate synapse (Llinás *et al.*, 1972), and in *Aplysia* ganglion cell bodies (Stinnakre and Tauc, 1973).

SIGNIFICANCE OF CALCIUM SPIKES

In nerve cells, Ca spikes may play a role in neurosecretion: calcium-dependent action potentials have been found in squid presynaptic endings (Katz and Miledi, 1967, 1969), in sympathetic ganglion cells (Koketsu *et al.*, 1959) and in crayfish neurosecretory cells (Iwasaki and Satow, 1971). Calcium ions are also necessary for the release of vasopressin in the pituitary (Douglas and Poisner, 1964).

In muscle cells from invertebrates, contraction often occurs without any regenerative action potentials. However, when action potentials do occur (with reduced intracellular Ca concentration or following treatment with procaine or TEA), they are Ca spikes (Fatt and Katz, 1953; Fatt and Ginsborg, 1958; Hagiwara and Naka, 1964; and Hagiwara *et al.*, 1964). Vertebrate muscle fibers, on the other hand, use sodium ions for the production of action potentials (Nastuk and Hodgkin, 1950). Reversal of ciliary beating in paramecia has been shown to result from an increased Ca conductance (Naitoh *et al.*, 1972).

In summary, it appears that Ca currents across excitable membranes occur in connection with EFFECTORS, i.e., secretory or contractile structures.

PROBLEMS

1. The transmembrane potential of a completely calcium-selective electrode is given by

$$E = \frac{RT}{2F} \ln \frac{[Ca]_o}{[Ca]_i}$$

What change in potential is produced by a ten-fold change in $[Ca]_o$? [Remember that $(RT/F) \ln A = 58 \log A$.]

2. What change in potential is produced by a two-fold change in $[Ca]_o$?

3. In Fatt and Ginsborg's treatment of bi-ionic potentials, there was no potassium in the external solution and negligible internal strontium. The currents for these ions were then given by

$$I_{Sr} = -4P_{Sr} \frac{EF^2}{RT} \frac{[Sr]_o \exp(-2EF/RT)}{1 - \exp(-2EF/RT)}$$

$$I_K = P_K \frac{EF^2}{RT} \frac{[K]_i}{1 - \exp(-EF/RT)}$$

At the peak of the action potential, they assumed that $I_{Sr} + I_K = 0$. Show that this leads to Equation 1.

4. In Equation 1, how negative must E be in order that Equation 3 is a reasonable approximation? (You might take as a criterion that 0.05 is very small compared with 1.)

5. In the data of Figure 2, when $[K]_i = 48$ mM, the spike overshoot (OS) varied with $[Ca]_o$ as follows:

$[Ca]_o$, mM	OS, mV
20	6.9
85	27.5
338	47.5

What is the approximate change in overshoot for a ten-fold change in $[Ca]_o$?

REFERENCES

Ashley, C. C., and Ridgway, E. B. (1970). On the relationship between membrane potential, calcium transient and tension in single barnacle muscle fibres, *J. Physiol.* **209**, 105–130.

Baker, P. F., Hodgkin, A. L., and Ridgway, E. B. (1971). Depolarization and calcium entry in squid giant axons, *J. Physiol.* **218**, 709–755.

Bülbring, E., and Kuriyama, H. (1963). Effects of changes in the external sodium and calcium concentrations on spontaneous electrical activity in smooth muscle of guinea pig *taenia coli, J. Physiol.* **166**, 29–58.

Douglas, W. W., and Poisner, A. M. (1964). Stimulus-secretion coupling in a neurosecretory organ: The role of calcium in the release of vasopressin from the neurohypophysis, *J. Physiol.* **172**, 1–18.

Fatt, P., and Ginsborg, B. L. (1958). The ionic requirements for the production of action potentials in crustacean muscle fibres, *J. Physiol.* **142**, 516–543.

Fatt, P., and Katz, B. (1953). The electrical properties of crustacean muscle fibres, *J. Physiol.* **120**, 171–204.

Geduldig, D., and Gruener, R. (1970). Voltage clamp of the *Aplysia* giant neurone: early sodium and calcium currents, *J. Physiol.* **211**, 217–244.

Geduldig, D., and Junge, D. (1968). Sodium and calcium components of action potentials in the *Aplysia* giant neurone, *J. Physiol.* **199**, 347–365.

Gerasimov, V. D., Kostyuk, T. G., and Maiskii, V. A. (1964). Excitability of the giant nerve cells of various lunged molluscs (*Helix pomatia, Limnea stagnalis* and *Planorbis corneus*) in solutions free from sodium ions, *Bull. Exp. Biol. Med.* **58**, 3–7 (translated from Russian in FASEB translation S65-4).

Hagiwara, S., Chichibu, S., and Naka, K. (1964). The effects of various ions on resting and spike potentials of barnacle muscle fibers, *J. Gen. Physiol.* **48**, 163–179.

Hagiwara, S., and Naka, K. (1964). The initiation of spike potential in barnacle muscle fibers under low intracellular Ca^{++}, *J. Gen. Physiol.* **48**, 141–162.

Irisawa, H., Shigeto, N., and Otani, M. (1967). Effect of Na^+ and Ca^{2+} on the excitation of the Mytilus (bivalve) heart muscle, *Comp. Biochem. Physiol.* **23**, 199–212.

Iwasaki, S., and Satow, Y. (1971). Sodium- and calcium-dependent spike potentials in the secretory neuron soma of the x-organ of the crayfish, *J. Gen. Physiol.* **57**, 216–238.

Junge, D. (1967). Multi-ionic action potentials in molluscan giant neurones, *Nature, Lond.* **215**, 546–548.

Katz, B., and Miledi, R. (1967). A study of synaptic transmission in the absence of nerve impulses, *J. Physiol.* **192**, 407–436.

Katz, B., and Miledi, R. (1969). Tetrodotoxin-resistant electric activity in presynaptic terminals, *J. Physiol.* **203**, 459–487.

Kerkut, G. A., and Gardner, D. R. (1967). The role of calcium ions in the action potentials of *Helix aspersa* neurones, *Comp. Biochem. Physiol.* **20**, 147–162.

Koketsu, K., Cerf, J. A., and Nishi, S. (1959). Further observations on the activity of frog spinal ganglion cells in sodium-free solutions, *J. Neurophysiol.* **22**, 693-703.

Llinás, R., Blinks, J. R., and Nicholson, C. (1972). Calcium transient in presynaptic terminal of squid giant synapse: detection with aequorin, *Science* **176**, 1127–1129.

Meves, H. (1966). Das Aktionspotential der Riesennervenzellen der Weinbergschnecke *Helix pomatia, Pflügers Arch. Ges. Physiol.* **289**, R10.

Meves, H., and Vogel, W. (1973). Calcium inward currents in internally perfused giant axons, *J. Physiol.* **235**, 225–265.

Miyazaki, S., Takahashi, K., and Tsuda, K. (1972). Calcium and sodium contributions to regenerative responses in the embryonic excitable cell membrane, *Science* **176**, 1441–1443.

Naitoh, Y., Eckert, R., and Friedman, K. (1972). A regenerative calcium response in *Paramecium, J. Exp. Biol.* **56,** 667–681.

Nastuk, W. L., and Hodgkin, A. L. (1950). The electrical activity of single muscle fibers, *J. Cell. Comp. Physiol.* **35,** 39–73.

Niedergerke, R., and Orkand, R. K. (1966a). The dual effect of calcium on the action potential of the frog's heart, *J. Physiol.* **184,** 291–311.

Niedergerke, R., and Orkand, R. K. (1966b). The dependence of the action potential of the frog's heart on the external and intracellular sodium concentration, *J. Physiol.* **184,** 312–334.

Sperelakis, N., Schneider, M. F., and Harris, E. J. (1967). Decreased K^+ conductance produced by Ba^{++} in frog sartorius fibers, *J. Gen. Physiol.* **50,** 1565–1583.

Stinnakre, J., and Tauc, L. (1973). Calcium influx in active *Aplysia* neurones detected by injected aequorin, *Nature New Biol.* **242,** 113–115.

Twarog, B. M. (1967). Excitation of *Mytilus* smooth muscle, *J. Physiol.* **192,** 857–868.

Wald, F. (1972). Ionic differences between somatic and axonal action potentials in snail giant neurones, *J. Physiol.* **220,** 267–281.

Watanabe, A., Tasaki, I., Singer, I., and Lerman, L. (1967). Effects of tetrodotoxin on excitability of squid giant axons in sodium-free media, *Science* **155,** 95–97.

9

METABOLIC PUMPS
AND MEMBRANE POTENTIALS

POSTTETANIC HYPERPOLARIZATION

The idea that part of the membrane potential of nerve and muscle cells is due to an active transport process is not new: as early as 1951, Hodgkin speculated on the existence of such an ELECTROGENIC, or potential-producing, pump. However, the presence of such a pump was difficult to prove or disprove in the squid axon, because its electrical effects were so small. Around the end of the 1950s, some examples of membrane-potential changes caused by pumps were found in connection with posttetanic hyperpolarization (PTH). This consisted of a hyperpolarization of several millivolts, lasting for up to a minute after the end of a TETANUS, or burst of activity. It was first seen in sympathetic nerve trunks and peripheral nerves (Ritchie and Straub, 1957; Connelly, 1959; Straub, 1961; and Holmes, 1962). The membrane potentials of the fibers were measured across a sucrose gap which isolated an active from an inactive region of membrane (see Chapter 2). A typical record of this behavior is shown in Figure 1, obtained with a cervical sympathetic trunk. The tetanus was produced by extracellular stimuli delivered to the trunk at 50 Hz for 5 sec. Immediately after the end of the tetanus, a hyperpolarization of about 3mV was observed; it decayed for several seconds. During the next few years, further examples of PTH were found in the crayfish stretch receptor (Nakajima and Takahashi, 1966), rabbit vagus nerve (Rang and Ritchie, 1968; and Den Hertog and Ritchie, 1969), and leech ganglion cells (Baylor and Nicholls, 1969). All had in common the property that a burst of action potentials was followed by several seconds of hyperpolarization below the previous resting potential.

113

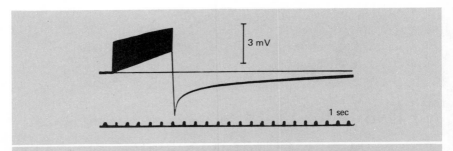

Posttetanic hyperpolarization in sympathetic nerve trunk. Black area caused by overlapping action potentials during tetanic (50-Hz) stimulation. (Holmes, 1962.)

ELECTROGENIC PUMPS AND BLOCKING AGENTS

Two known mechanisms which could give rise to the phenomenon of PTH are (1) stimulation of an electrogenic pump and (2) an increased conductance to ions with an equilibrium potential more negative than resting (probably K^+). The pump mechanism involves the active transport of Na^+ and K^+ which maintains the asymmetric distribution of these ions across excitable membranes (see Chapter 4). Because less potassium is pumped in than sodium is pumped out, a net deficit of positive ions is produced on the inner side of the membrane and a hyperpolarization results.

All the above examples of PTH have been shown to result from stimulation of electrogenic pump activity. Some of the evidence used in support of this conclusion included (1) that the PTH was reduced by lowering external K concentration. This would be expected with a coupled pump, where external potassium is necessary for the Na–K exchange to take place. It would not be expected for a K-conductance mechanism, because lowering external potassium should make the potassium equilibrium potential more negative and *increase* the PTH. (2) The PTH was reduced or blocked by replacing external Na with Li. Evidently, lithium leaked into the cells but could not be pumped out by the Na pump. (3) Agents which interfered with the production of adenosine triphosphate (ATP) (iodoacetate, azide, cyanide, and dinitrophenol) also blocked the PTH. (4) The PTH was reduced by ouabain or strophanthidin, which do not interfere with the production of ATP but block the active transport process in the membrane. (5) Cooling the preparation to 5 to 10°C strongly reduced the PTH, presumably by inhibiting the ATPase involved in the Na–K exchange process. The effect of cooling was much larger than would be expected for a K-conductance mechanism. For instance, a Goldman-Hodgkin-Katz membrane potential (Equation 18, Chapter 4) would only vary 4% upon cooling from 22 to 10°C.

Although all the above examples of PTH were shown to result from stimulation of electrogenic pumps, it should be mentioned that a K-conductance increase produced the PTH seen in the node of Ranvier (Meves, 1961), phrenic motor nerve terminals (Gage and Hubbard, 1966), and *Aplysia* ganglion cells (Brodwick and Junge, 1972). However, this chapter is directed to the properties of electrogenic pumps, and not PTH *per se*.

STIMULATION OF ELECTROGENIC PUMPS

Hodgkin and Keynes (1956) showed that injection of sodium ions into squid axons stimulated active extrusion of sodium by the pump. The extrusion of labeled Na^+ was proportional to the intracellular Na concentration, suggesting that increasing the substrate accelerated the pump by mass action. In addition, Na injection produced a small (1.6-mV) hyperpolarization.

The first demonstration of an electrogenic pump in muscle (Kernan, 1962) also involved increasing the intracellular Na concentration: frog skeletal muscle was soaked in K-free solutions at low temperatures; this blocked active Na pumping and caused "loading" of the fibers with sodium. When the skeletal muscle was transferred to a recovery solution at room temperature with normal K concentration, the membrane potentials became about 11 mV more negative than the calculated E_K. (This result stimulated some controversy because many investigators at that time considered the Na–K pump in muscle to be completely coupled, and therefore neutral.) Then in 1965, Mullins and Awad showed that Na-loaded muscles underwent a large hyperpolarization upon being warmed or transferred from a K-free to a high-K medium, while muscles with a normal $[Na]_i$ did not. The next year it was shown (Adrian and Slayman, 1966) that Na-loaded muscle could attain membrane potentials 20 mV more negative than E_K and that this hyperpolarization was abolished by ouabain. By 1966, the idea of an electrogenic pump in frog muscle was firmly established.

About this time, sodium injection was also shown to cause hyperpolarization in nerve cell bodies in snails by stimulation of an electrogenic pump (Kerkut and Thomas, 1965). The data of one such experiment are shown in Figure 2. At time zero, an electrode containing concentrated potassium acetate was inserted into the cell body;

2

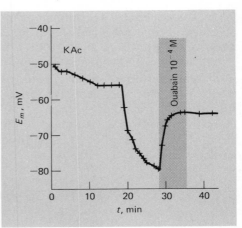

Effect of ouabain on Na-induced hyperpolarization in snail neurons. At time 0, potassium acetate injected. After about 20 min, sodium acetate injected. Ouabain added to external solution after injection of Na. (Kerkut and Thomas, 1965.)

diffusion of KAc into the cell caused a slight increase in the resting potential, as would be expected for a K-permeable membrane. Twenty minutes later, an electrode filled with concentrated sodium acetate was inserted, and the membrane potential rapidly became 25 mV more negative. The effect of Na injection was offset by ouabain, indicating that the sodium injection had produced the hyperpolarization by means of a pump. Nakajima and Takahashi (1966) also observed a long-lasting hyperpolarization of the crayfish stretch receptor after electrophoretic injection of sodium citrate. Other studies have been performed with Na injection in snail neurons and will be discussed in connection with measurement of pump currents.

For a number of years, there has been some dispute about whether an electrogenic pump operates in the squid axon, perhaps because the changes in membrane potential produced by the pump are smaller than those in nerve cell bodies. Recently, however, De Weer and Geduldig (1973) have carried out careful measurements of squid axon membrane potentials using a low-drift recording system. A depolarization of 1.4 mV was produced by application of strophanthidin to the axons; the depolarization was larger in Na-loaded axons and was blocked by cyanide. This observation appears to support the idea that there are no truly neutral (completely coupled) Na–K pumps.

As mentioned in connection with sodium-loaded muscle fibers, warming can stimulate the action of electrogenic pumps. Some investigators have taken the approach of studying pumps in nerve cells principally by warming. For instance, Carpenter and Alving (1968) observed the behavior of the *Aplysia* giant (R2) cell shown in Figure 3. In part A, the cell was warmed from 5 to 21°C, and a hyperpolarization of 9 mV occurred. In part B, the same procedure was repeated in the presence of ouabain, producing a depolarization with superimposed action potentials instead of a hyperpolarization. The hyperpolarization in part A was thus apparently caused by an electrogenic pump, and it was suggested that the depolarization in part B resulted from an increase in membrane Na conductance upon warming. Subsequently, Marchiafava (1970) studied the effects of warming on this cell in the presence of ouabain and concluded that warming increased both the resting g_{Na} and g_K, but increased g_{Na} more than g_K; Carpenter (1970) reached the same conclusion.

Gorman and Marmor (1970) carried out a quantitative study of the effects of temperature and external potassium concentration on membrane potentials in *Anisodoris* neurons. The result of one of their experiments is shown in Figure 4. The open circles show the variation of exp (EF/RT) with external K concentration at 5°C when the electrogenic pump is blocked. The straightness of this line is consistent with the constant-field expression for membrane potential (Equation 20, Chapter 4). The solid dots show the variation of exp (EF/RT) with $[K]_o$ at 18°C. Near $[K]_o = 0$ the pump is inhibited, and the potential is near that at 5°C. From about 5 to 50-mM $[K]_o$, the pump is activated, and the potential is always more negative than that predicted by the constant-field equation. At higher values of $[K]_o$, the pump component of potential is small compared with that due to the passive ion conductances, and the curves at 5 and 18°C overlap. Thus, changes in ambient temperature do exert a strong influence on the metabolically derived component of potential. However, this method of controlling the pump action is not as specific as the use of ouabain or stroph-

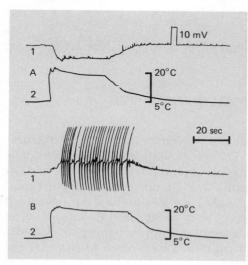

Effect of ouabain on the warming response of an *Aplysia* nerve cell. Top traces in A and B show membrane potential, bottom traces temperature near cell. Part A: Cell warmed from 5 to 21°C. Part B: Same warming with ouabain in external solution. (Carpenter and Alving, 1968.)

3

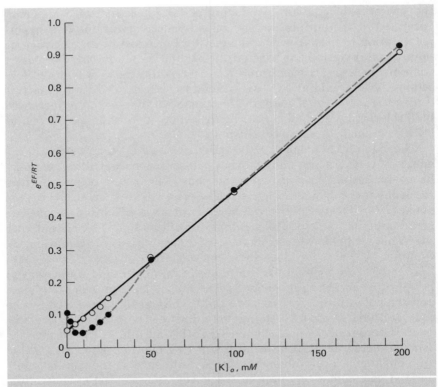

4

Variation of $e^{EF/RT}$ with external K concentration in mollusc neuron (E, membrane potential). (Gorman and Marmor, 1970.)

anthidin because of the additional effect of temperature on the ionic conductances.

MEASUREMENTS OF PUMP CURRENTS WITH VOLTAGE CLAMP

The first observation of sodium-pump currents using the voltage-clamp technique was made by Thomas (1969), in snail neurons. The clamp circuit was designed with a very long time constant, so that only average membrane potential was clamped and not individual action potentials. Currents measured by this method are shown in Figure 5. The start and end of injection of sodium acetate into a neuron cell body are indicated by downward- and upward-going marks near the start of each trace. Longer injections were used to increase the amount of injected sodium. Outward current, shown as a downward deflection, increased with increasing Na injected. The remaining outward currents after the end of the injections declined with an average time constant of 4.4 min. The pump currents were blocked by ouabain or K-free external solutions. This investigator also used intracellular Na-sensitive microelectrodes to measure internal sodium concentration and found that the pump current was proportional to the excess of $[Na]_i$ over the normal level.

Subsequently, Kostyuk *et al.* (1972) carried out some further voltage-clamp studies of pump currents in snail neurons. They examined the effect of membrane potential on the currents in response to Na injection and found that the outward current decreased with increasing hyperpolarization. They attributed this behavior to a potential-dependent active K current and a potential-independent Na current. Because the effect of the pump in the unclamped membrane is one of hyperpolarization, the potential dependence of the pump may

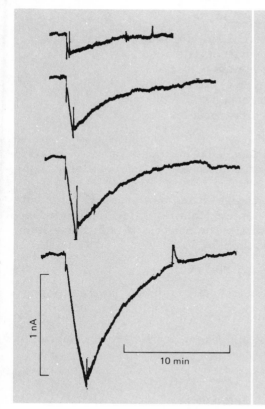

1 nA

10 min

Sodium-pump currents measured with voltage clamp in snail neurons. (Thomas, 1969.)

5

act as a negative feedback. Some important questions remain to be studied regarding pump currents, but these observations provide the bases for a conceptual model of the pump mechanism.

CHARACTERIZATION OF THE PUMP

Several different models of varying complexity have been put forth to summarize our knowledge about electrogenic active transport (Moreton, 1969; Rapoport, 1970; and Martirosov and Mikayelyan, 1970). However, the model of Mullins and Noda (1963) is the simplest that includes the effects of varying the Na–K coupling ratio of the pump.

The net passive flux of an ion into or out of the cell can be derived as

$$\overline{m} = P([C]_o - [C]_i e^{EF/RT})\, f(E) \tag{1}$$

where

$$
\begin{aligned}
P &= \text{permeability of the ion} \\
[C]_o &= \text{external concentration} \\
[C]_i &= \text{internal concentration} \\
E &= \text{membrane potential} \\
R &= \text{universal gas constant} \\
T &= \text{temperature, } °K \\
F &= \text{Faraday constant}
\end{aligned}
$$

The function $f(E)$ describes the potential dependence of the passive flux and is the same for any univalent ion. It is equal to $EF/[RT(1 - e^{EF/RT})]$ for the constant-field case but need not be stated in deriving this model.

It is assumed that only passive fluxes of Na and K contribute to the resting membrane potential (chloride being distributed according to the resulting potential). The net passive influx of sodium, $\overline{m}_{Na}$, is balanced by an active efflux, $\overline{p}_{Na}$:

$$\overline{m}_{Na} + \overline{p}_{Na} = 0 \tag{2}$$

Similarly, for the passive efflux and active influx of potassium

$$\overline{m}_K + \overline{p}_K = 0 \tag{3}$$

If r is the number of Na ions pumped out for each K ion pumped in, then

$$r\overline{p}_K + \overline{p}_{Na} = 0 \tag{4}$$

Substituting from Equations 2 and 3 gives

$$r\overline{m}_K + \overline{m}_{Na} = 0 \tag{5}$$

From Equation 1,

$$rP_K ([K]_o - [K]_i e^{EF/RT}) f(E) + P_{Na} ([Na]_o - [Na]_i e^{EF/RT}) f(E) = 0 \quad (6)$$

This may be rewritten as

$$E = \frac{RT}{F} \ln \frac{rP_K [K]_o + P_{Na} [Na]_o}{rP_K [K]_i + P_{Na} [Na]_i} \quad (7)$$

If $r = 1$, that is, the outward flux of sodium is exactly equal to the inward flux of potassium, then Equation 7 becomes the Goldman-Hodgkin-Katz expression for the potential across a membrane permeable only to sodium and potassium. If r becomes infinite, the condition where no potassium ions are pumped (zero net flux of K), then the potential becomes that of a potassium electrode. For intermediate cases, such as $r = 3$, the predicted potential is closer to the potassium equilibrium potential than for $r = 1$. This model is thus able to incorporate the observation of Kostyuk et al. (1972) that the coupling ratio r is potential-dependent. Of course, as more facts are uncovered concerning the behavior of electrogenic pumps, the model will have to be modified to include them. But it provides a basis for understanding metabolically dependent membrane potentials, just as the Goldman-Hodgkin-Katz theory does for the ionic dependencies of membrane potentials.

PROBLEMS

1. Using Equation 7 with $r = 1$ (completely coupled pump), calculate the expected resting potential for a muscle cell with

$$[Na]_o = 115 \text{ m}M$$
$$[Na]_i = 10 \text{ m}M$$
$$[K]_o = 2.5 \text{ m}M$$
$$[K]_i = 140 \text{ m}M$$
$$P_{Na}/P_K = 0.015$$

2. What is the value of the resting potential if r is increased to the typical value in muscle of 1.5?

3. Thus, what is the pump contribution to the resting potential in this cell?

4. An alternative model of the electrogenic pump uses the parallel-conductance formulation of the sodium and potassium currents. Thus, in Equation 5

$$r\overline{m}_K + \overline{m}_{Na} = 0$$

The passive fluxes of potassium and sodium are described by

$$\overline{m}_K F = I_K = g_K (E - E_K)$$
$$\overline{m}_{Na} F = I_{Na} = g_{Na} (E - E_{Na})$$

Where F = the Faraday constant, 96,500 coulombs/mol. From these relationships, find the membrane potential, E, in terms of g_K, g_{Na}, E_K, E_{Na}, and r.

5. In the parallel-conductance model of the electrogenic pump in Problem 4, the appropriate parameters for a muscle fiber membrane are

$$E_K = -101.4 \text{ mV}$$
$$E_{Na} = 61.5 \text{ mV}$$
$$g_K = 200 \ \mu\text{mho/cm}^2$$
$$g_{Na} = 17.6 \ \mu\text{mho/cm}^2$$

Calculate the membrane potential when $r = 1$.

6. Find the potential as in Problem 5 when $r = 1.5$.

REFERENCES

Adrian, R. H., and Slayman, C. L. (1966). Membrane potential and conductance during transport of sodium, potassium and rubidium in frog muscle, *J. Physiol.* **184**, 970–1014.

Baylor, D. A., and Nicholls, J. G. (1969). After-effects of nerve impulses on signalling in the central nervous system of the leech, *J. Physiol.* **203**, 571–589.

Brodwick, M. S., and Junge, D. (1972). Post-stimulus hyperpolarization and slow potassium conductance increase in *Aplysia* giant neurone, *J. Physiol.* **223**, 549–570.

Carpenter, D. O. (1970). Membrane potential produced directly by the Na+ pump in *Aplysia* neurons, *Comp. Biochem. Physiol.* **35**, 371–385.

Carpenter, D. O., and Alving, B. O. (1968). A contribution of an electrogenic Na+ pump to membrane potential in *Aplysia* neurons, *J. Gen. Physiol.* **52**, 1–21.

Connelly, C. M. (1959). Recovery processes and metabolism of nerve, *Rev. Mod. Phys.* **31**, 475–484.

De Weer, P., and Geduldig, D. (1973). Electrogenic sodium pump in squid giant axon, *Science* **179**, 1326–1328.

Den Hertog, A., and Ritchie, J. M. (1969). A comparison of the effect of temperature, metabolic inhibitors and of ouabain on the electrogenic component of the sodium pump in mammalian non-myelinated nerve fibres, *J. Physiol.* **204**, 523–538.

Gage, P. W., and Hubbard, J. I. (1966). The origin of the post-tetanic hyperpolarization of mammalian motor nerve terminals, *J. Physiol.* **184**, 335–352.

Gorman, A. L. F., and Marmor, M. F. (1970). Contributions of the sodium pump and ionic gradients to the membrane potential of a molluscan neurone, *J. Physiol.* **210**, 897–917.

Hodgkin, A. L. (1951). The ionic basis of electrical activity in nerve and muscle, *Biol. Rev.* **26**, 339–409.

Hodgkin, A. L., and Keynes, R. D. (1956). Experiments on the injection of substances into squid giant axons by means of a microsyringe, *J. Physiol.* **131**, 592–616.

Holmes, O. (1962). Effects of pH, changes in potassium concentration and metabolic inhibitors on the after-potentials of mammalian non-medullated nerve fibres, *Archs Int. Physiol.* **70**, 211–245.

Kerkut, G. A., and Thomas, R. C. (1965). An electrogenic sodium pump in snail nerve cells, *Comp. Biochem. Physiol.* **14**, 167–183.

Kernan, R. P. (1962). Membrane potential changes during sodium transport in frog sartorius muscle. *Nature, Lond.* **193**, 986–987.

Kostyuk, P. G., Krishtal, O. A., and Pidoplichko, V. I. (1972). Potential-dependent membrane current during the active transport of ions in snail neurones, *J. Physiol.* **226**, 373–392.

Marchiafava, P. L. (1970). The effect of temperature change on membrane potential and conductance in *Aplysia* giant nerve cell, *Comp. Biochem. Physiol.* **34**, 847–852.

Martirosov, S. M., and Mikayelyan, L. G. (1970). Ion exchange in electrogenic active transport of ions, *Biofizika* **15**, 104–111.

Meves, H. (1961). Die Nachpotentiale isolierter markhaltiger Nervenfasern des Frosches bei tetanischer Reizung. *Pflüger's Arch. ges. Physiol.* **272**, 336–359.

Moreton, R. B. (1969). An investigation of the electrogenic sodium pump in snail neurones, using the constant-field theory, *J. Exp. Biol.* **51**, 181–201.

Mullins, L. J., and Awad, M. Z. (1965). The control of the membrane potential of muscle fibers by the sodium pump, *J. Gen. Physiol.* **48**, 761–775.

Mullins, L. J., and Noda, K. (1963). The influence of sodium-free solutions on the membrane potential of frog muscle fibers, *J. Gen. Physiol.* **47**, 117–132.

Nakajima, S., and Takahashi, K. (1966). Post-tetanic hyperpolarization and electrogenic Na pump in stretch receptor neurone of crayfish, *J. Physiol.* **187**, 105–127.

Rang, H. P., and Ritchie, J. M. (1968). On the electrogenic sodium pump in mammalian non-myelinated nerve fibres and its activation by various external cations, *J. Physiol.* **196**, 183–221.

Rapoport, S. I. (1970). The sodium-potassium exchange pump: relation of metabolism to electrical properties of the cell: I. Theory, *Biophysic. J.* **10**, 246–259.

Ritchie, J. M., and Straub, R. W. (1957). The hyperpolarization which follows activity in mammalian non-medullated fibres, *J. Physiol.* **136**, 80–97.

Straub, R. W. (1961). On the mechanism of post-tetanic hyperpolarization in myelinated nerve fibres from the frog, *J. Physiol.* **159**, 19–20P.

Thomas, R. C. (1969). Membrane current and intracellular sodium changes in a snail neurone during extrusion of injected sodium, *J. Physiol.* **201**, 495–514.

10

NEW DIRECTIONS

The development of the field of nerve and muscle biophysics has at times been altered by radical new approaches, such as the sucrose gap, microelectrode, and voltage clamp. Although these innovative methods seemed strange when introduced, they have now become completely accepted and are always included in textbooks such as this one. In retrospect, it is possible to see what a wealth of information was yielded by each technique. There are, at present, several promising new lines of study of excitable membranes utilizing as yet untried methods. These approaches are deserving of mention, although it is too soon to know how important each will become.

INSIDE-OUT AXONS

Spyropoulos (1972) has presented a method of turning axons from the large squid *Loligo vulgaris* inside-out and examining the electrical properties of the everted axons. This approach permits removing more of the axoplasm than is possible with the internal-perfusion technique, in order to study whether the membrane itself has *asymmetric* permeability properties. If the membrane is symmetric, then the resting and active properties must be merely a result of the normal ion distributions, and not the result of any difference between the inside and outside of the membrane itself. In this case, the everted axon should behave just like a normal one, if the internal and external ion concentrations are normal. On the other hand, if the membrane is asymmetric, then it should make a difference which side is outside. Figure 1 shows an action potential recorded from an everted axon with seawater inside and isotonic potassium fluoride outside. The resting potential was greater than $+53$ mV, and the spike overshoot was at least -50 mV. The everted axon could work very well with the opposite ion distribution from normal. However, when the external solution was changed to seawater and the internal to KF, the resting potential fell to about +20 mV, and no action potentials could be produced. This result indicated that the membrane itself was asymmetric, just as had been suggested by the earlier perfusion studies (Baker *et al.*, 1962). Spyropoulos was also able to show a great prolongation of the spike when tetraethylammonium was applied outside the everted axons, but no effect when it was applied internally. This confirmed the results of Armstrong and Binstock (1965) on normal axons and indicated that the everted axon was functioning in a normal manner. The fact that these results are the same as found for perfused right-side-out axons means that (1) the perfusion technique is a valid means for controlling the ionic medium on the axoplasmic side of the membrane, and (2) either the membrane itself, or the membrane with approximately 1 μm of attached axoplasm, is an inherently asymmetric structure.

BIREFRINGENCE

During spike activity, excitable structures show changes in the optical properties of the birefringence and light scattering. BIREFRINGENCE

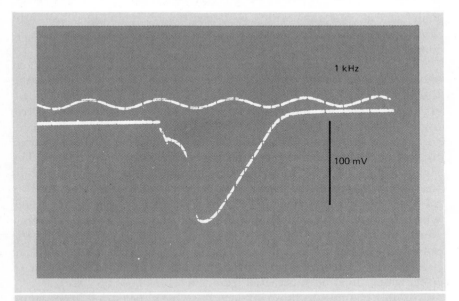

1

Action potential from inside-out squid axon with seawater inside and isotonic KF outside. (Spyropoulos, 1972.)

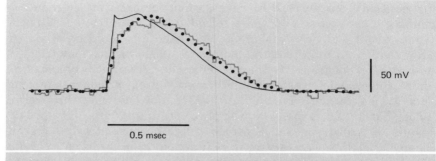

50 mV

0.5 msec

2

Birefringence change during activity in eel electric organ. Averaged change in light intensity (stepped line) compared with membrane potential (solid line) and potential delayed by 55 μsec (solid dots). (Hille, 1970.)

is a property of materials which have different indices of refraction, depending on the angle at which light impinges on them; calcite crystals are a common example. When plane-polarized light passes through a birefringent material, the emergent light has a component at 90° to the direction of polarization of the incident light. This component can thus pass through an ANALYZER oriented at 90° to the plane of the POLARIZER which produced the incident light. The light which passes through the crossed polarizer and analyzer can then be measured with a photomultiplier.

The changes in birefringence produced in axons or electric-organ slices from electric eels are small but can be seen clearly by averaging the light changes produced by hundreds of electric stimuli. The birefringence change due to electric stimulation of a slice of electric organ is shown in Figure 2. The solid line is the action potential recorded in a single cell of the electric organ (the eel's "jolt" is produced by adding thousands of these single-cell potentials in series). The stepped line shows the birefringence change, which closely resembles the membrane-potential record. If the potential record is delayed by 55 μsec (with some smoothing), then the dotted line is obtained, which is practically superimposable with the light record. In other words, it appears that the birefringence change is produced by the potential change after a 37-μsec delay for an intervening coupling process (the light-measuring circuit already introduced a delay of 18 μsec, which must be subtracted from the total lag of 55 μsec).

Similar changes in birefringence have been measured in the squid axon. The changes during the action potential were greatest at the edge of the axon (viewed from one side) and least at the center. In addition, the light signal was the same after removal of 95% of the axoplasm and perfusing with a solution of potassium fluoride, which maintained excitability (Cohen *et al.*, 1968). These observations suggest that the birefringence change is localized at or near the axon membrane, rather than in the axoplasm.

Such changes might occur from either of two mechanisms: the KERR EFFECT, resulting from realignment of molecules in the membrane; or compression of the membrane by the electrostatic force of the charge separation. Cohen *et al.* (1969) have calculated that either of these effects could be large enough with the known potential shifts (about 100 mV) to account for the observed changes in birefringence. What is clear is that this optical property is strongly tied to membrane potential and that it probably arises from physical alterations of the membrane during excitation.

LIGHT SCATTERING

If a section of a nerve or an electric organ is placed in the path of a narrow light beam, a wide portion of the object becomes luminous, not just the spot struck by the beam. This diffusion of the light beam is called SCATTERING, and the amount of light given off at any angle to the beam may be measured with a photosensitive device. In 1952, Bryant and Tobias showed that the light scattered at 90° to the incident beam by a section of crab nerve decreased during and after a burst of action potentials. Subsequently, Cohen and his co-workers (1968, 1969) have shown changes in scattering during a single action potential in the eel electric organ and squid axon. These variations are quite small and can only be detected with signal-averaging techniques.

The data in Figure 3 were obtained from a voltage-clamped squid axon and show the low-angle scattering change, near the direction of the incident light beam. The top trace is the scattered light at about 35° from the beam, the middle trace is the membrane potential, and the bottom trace is the voltage-clamp current. At the left of the figure, a hyperpolarizing command pulse was applied with little effect on the light signal or current record. Then a depolarizing pulse was applied, and the familiar inward-then-outward pattern of clamp current was seen. At the same time, the scattered light showed a rapid decrease followed by a longer-lasting increase. By analyzing voltage-clamp records, these authors concluded that the scattered light in the squid axon was partially dependent on both membrane potential and current; i.e. that no simple interpretation of the optical changes was possible.

Some possible structural changes in the excitable tissues which

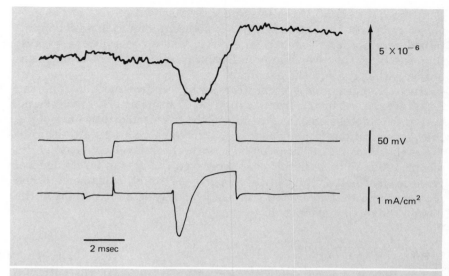

5×10^{-6}

50 mV

1 mA/cm^2

2 msec

Low-angle light scattering change during activity in voltage-clamped squid axon. Light calibration shows change in scattering divided by resting scattering. (Cohen, 1970.)

could have caused the scattering effects include (1) movements of ions, possibly changing the refractive index in or around the excitable membranes, (2) conformational changes in molecules of the membranes or cytoplasm, (3) macromolecular structure changes, or (4) swelling or shrinking of cellular organelles. Cohen (1970) points out that the time course of the *low-angle* scattering change in the squid axon is about the same as that expected for K accumulation in the space immediately around the axon, following a train of impulses or a single action potential. This scattering change may result from local alterations of the index of refraction, due to K accumulation. The *right-angle* scattering does not follow this time course and will require a more complicated explanation.

FLUORESCENCE

In addition to the measurement of changes in birefringence and light scattering, it is also possible to stain axons with fluorescent dyes and record changes in fluorescence during excitation. These dyes emit light of a different wavelength than the ultraviolet light which is used to excite them. With an absorption filter, the incident light may be excluded from the recording photosensitive device. As with the other

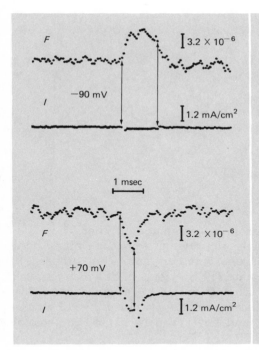

Fluorescence changes of ANS-injected squid axon in response to voltage-clamped command pulses. F, averaged fluorescent light emission. I, clamp currents. Light calibration shows average fluorescence divided by that obtained with a 70 mV depolarizing clamp pulse. (Conti *et al.*, 1971.)

optical techniques mentioned, the signals are extremely small, and averaging techniques must be used. Figure 4 shows the light recorded from a voltage-clamped squid axon injected with the fluorescent dye 1-analinonaphthalene-8-sulfonate (ANS), taken from Conti *et al.* (1971). The top trace in each part shows the average of 8,192 light responses, and the bottom trace the voltage-clamp current. A 90-mV hyperpolarizing command pulse (top part) produced little change in the current and a rapid increase in fluorescence intensity, which reached a maximum within 1 msec. A 70-mV depolarization (bottom part) elicited an inward current and a *decrease* in fluorescence with a much longer time constant than the response to a hyperpolarization. The amplitude of the increase and decrease in fluorescence were comparable, even though the depolarization was only 78% as large as the hyperpolarization. No satisfactory explanation is available for this difference.

ANS is a *hydrophobic* probe; that is, it only fluoresces in the presence of nonpolar (lipid) solvents. Thus, changes in fluorescence of squid axons may reflect a change in hydrophobicity (lipid solubility) of the axon membrane during excitation. Another property of the light from ANS-injected axons is that it is polarized in the direction of the axon. Tasaki and co-workers (1972, 1973) have shown this by inserting

5

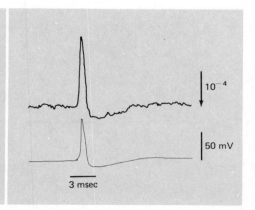

Increase in light absorption during activity in squid axon stained with mero-cyanine dye. Top trace: Transmitted light (increasing downward) at 570 nm. Light calibration shows change in transmitted light divided by resting transmission. Bottom trace: Intracellularly recorded action potential. (Ross *et al.*, 1974.)

10^{-4}

50 mV

3 msec

polarizing filters between the axon and the photomultiplier. The implication is that the oscillating molecules giving rise to the emitted light are oriented parallel to the axon. Another fluorescent probe, 2-*p*-toluidinyl-naphthalene-6-sulfonate (TNS), yields light which is much more strongly polarized than that from ANS (Tasaki *et al.*, 1973). Fluorescent dyes may also be incorporated into whole nerves, by soaking rather than by injecting, and have even been used in skeletal muscle (Carnay and Barry, 1969).

LIGHT ABSORPTION

Several dyes have recently been identified which when applied to nerve membranes show an increased absorption of transmitted light during the action potential (Ross *et al.*, 1974). This is illustrated in Figure 5 for the dye merocyanine applied to the squid axon. (The signal-to-noise ratio with this method is about 20 times as large as with fluorescence measurements.) The bottom trace shows an action potential, and the top trace the simultaneous *decrease* in transmitted light at 570 nm (increased absorption). No signal could be recorded with the light source turned off, indicating that the optical recording was not due to electric coupling between the action potential and the light-recording system. Also, no signal could be recorded in unstained axons, indicating that the signal was not produced by some optical effect, such as light scattering. As shown in Figure 5, the change in light absorption during excitation closely follows the axon membrane potential, although the mechanism for the absorption change is unknown.

It is possible that staining with merocyanine-like dyes will permit the monitoring with optical techniques (e.g., direct videotaping) of activity in individual nerve cells in a complex structure such as a ganglion.

BATRACHOTOXIN

After the discovery of the highly specific blocking effect of tetrodotoxin (TTX) on the active sodium conductance (Narahashi *et al.*, 1964), it is not difficult to defend the use of neurotoxins as biophysical tools. Recently, a highly specific toxin, batrachotoxin (BTX), has been purified from the skin secretion of the Colombian arrow poison frog, *Phyllobates aurotaenia*. Batrachotoxin causes a reversible increase in sodium conductance in the *resting* membrane, which completely depolarizes the axons to which it is applied. Furthermore, the action of BTX is completely blocked by TTX. This effect is illustrated in Figure 6, obtained in the squid axon. The curve shows the resting potential as a function of time, as various external and internal solutions were applied to the perfused axon. First the external sodium concentration was reduced to 1 mM causing a slight hyperpolarization. Addition of 1-mM TTX in low- or normal-sodium external solution had little additional effect. When 0.55-mM BTX was added to the internal solution,

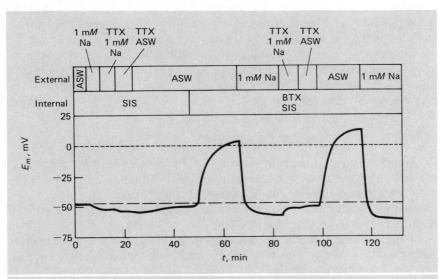

6

Depolarizing effect of batrachotoxin on squid axon. Changes in external solution shown in top row; internal solution in second row. ASW, artificial seawater. SIS, standard internal solution. (Narahashi *et al.*, 1971.)

a large depolarization developed over a period of about 20 min and eventually reversed the membrane potential. The dependence of this depolarization on sodium was shown by the next procedure: reducing the external sodium concentration to 1 mM with BTX inside the axon caused a rapid recovery of the resting potential. When TTX was added to the external solution with 1-mM Na$^+$, little effect was seen. Restoration of Na$^+$ to the normal level in the presence of TTX also had no effect, showing that tetrodotoxin had blocked the depolarizing action of BTX. When TTX was removed from the external solution, the BTX-induced depolarization was again seen in full force. Reduction of external sodium to 1 mM again cancelled the BTX effect. Thus, BTX apparently causes a large increase in the resting, or LEAKAGE, sodium permeability. The hyperpolarization produced by TTX when applied alone and the block of the BTX effect by TTX imply that the resting sodium channel is sensitive to TTX.

Interestingly enough, the *active* Na channel is not much affected by BTX: in the BTX-depolarized axon, if the membrane potential is restored to a normal resting level by artificial inward current, then normal-appearing action potentials may be produced.

Narahashi *et al.* (1971) have verified these observations with voltage-clamp methods, and Albuquerque *et al.* (1973) showed that the action of BTX on the resting sodium channel is unchanged if lithium is substituted for sodium in the external solution.

The effect of batrachotoxin is irreversible, that is, not removed by washing the BTX out of the external solution. Whether BTX acts as a charge carrier in the membrane or modifies a preexisting carrier or channel is not certain.

ACTION POTENTIALS IN ARTIFICIAL MEMBRANES

Although somewhat outside the scope of this book, the field of study of artificial membranes is developing rapidly and has produced some interesting suggestions about biological membranes. Among these is the demonstration that all-or-none action potentials can be produced in an artificial system. Mueller and Rudin (1968a) were able to form lipid bilayers of controlled composition between different electrolyte solutions and "dope" the resulting membranes with various current carriers, or IONOPHORES.

In one case, the membrane consisted of 4% oxidized cholesterol in decane, and the antibiotic alamethicin and the protein protamine were added to confer excitability. The electrolyte on one side was 100-mM NaCl, and on the other side 100-mM NaCl plus 40-mM KCl. An action potential produced by this system is shown in Figure 7. The

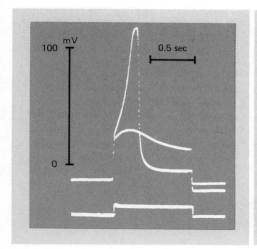

Action potential recorded in artificial membrane. Top trace: Potential. Bottom trace: Current. (Mueller and Rudin, 1968a.)

7

bottom trace shows applied currents, and the top traces the potential changes in response to subthreshold and suprathreshold stimuli. The action potential had an amplitude of more than 100 mV but was quite slow by neurophysiological standards (about 0.25-sec duration). Nevertheless, by using precise conditions of membrane and solution composition, these authors were able to produce regenerative action potentials in a completely controlled artificial system.

Mueller and Rudin (1968b) had also been able to obtain action potentials in bilayers treated with a proteinaceous bacterial extract called EXCITABILITY INDUCING MATERIAL, or EIM. However, that system was not completely determined because the structure of EIM was unknown.

OTHER APPLICATIONS OF IONOPHORES WITH KNOWN PROPERTIES

Finally, in this brief catalog of new approaches it is interesting to speculate on the possibilities of some ionophores which have been studied extensively in artificial membranes, but studied only to a limited extent in living systems. The macrotetralide antibiotic nonactin confers a potassium permeability upon lipid membranes, as shown in Figure 8. The lipid bilayer is made of glycerol dioleate and separates two solutions containing identical concentrations of KCl. The abscissa shows the concentration of nonactin in the bilayer, and the ordinate is membrane conductance. The KCl concentration on each side of the membrane is shown as a variable parameter. It can be seen that a 1,000-fold increase in nonactin concentration increases the K con-

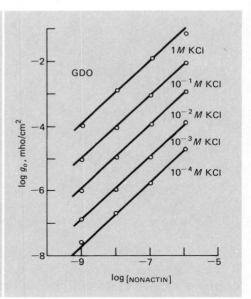

8

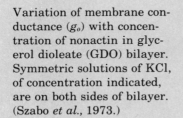

Variation of membrane conductance (g_o) with concentration of nonactin in glycerol dioleate (GDO) bilayer. Symmetric solutions of KCl, of concentration indicated, are on both sides of bilayer. (Szabo *et al.*, 1973.)

ductance 1,000 fold over the entire range of KCl concentration. Increased membrane conductance for all of the alkali cations, Li+, Na+, Cs+, Rb+, and K+, is produced by this and other antibiotics, such as valinomycin and gramicidin (Szabo *et al.*, 1973).

The artificial-membrane studies may lead in any of at least three interesting directions: (1) They provide model systems whose properties may be controlled more completely than those of living membranes. The models can suggest how the living systems *could* work. (2) By applying the ionophores to living membranes, we may be able to alter the behavior of the membranes in a new and useful way. For example, if the behavior of ionophores in living membranes is similar to that in artificial systems, it will be more convincing that the model systems are like the real ones. (3) It may become possible to alter the internal ion concentrations in nerve cell bodies and muscle fibers by adding suitable ionophores. If the membrane could be made highly permeable to sodium for a brief period, the cell could be loaded with sodium in a high-Na external solution. Alternatively, the cell could be made highly permeable to potassium, placed in a K-free medium, and depleted of internal potassium. The use of ionophores to change alkali cation concentrations inside cells is still speculative. However, a highly conductive *calcium* ionophore has already been used for just this purpose in secretory cells (Foreman *et al.*, 1973) and starfish eggs (Steinhardt *et al.*, 1974).

It is almost a foregone conclusion that by the time of publication of this book many other new and potentially fruitful approaches to the study of excitable membranes will be under test. Perhaps one of those outlined above or a method not yet thought of will open up a golden new vein of information. It is clear what a debt we owe to those careful and assiduous workers who provided such a useful foundation for the field up to this point. From now on, the challenge will be to generate new ways of "seeing" and to follow them out as thoroughly as have those who have gone before.

REFERENCES

Albuquerque, E. X., Seyama, I., and Narahashi, T. (1973). Characterization of batrachotoxin-induced depolarization of the squid giant axons, *J. Pharm. Exp. Ther.* **184,** 308–314.

Armstrong, C. M., and Binstock, L. (1965). Anomalous rectification in the squid giant axon injected with tetraethylammonium chloride, *J. Gen. Physiol.* **48,** 859–872.

Baker, P. F., Hodgkin, A. L., and Shaw, T. I. (1962). The effects of changes in internal ionic concentrations on the electrical properties of perfused giant axons, *J. Physiol.* **164,** 355–374.

Bryant, S. H., and Tobias, J. M. (1952). Changes in light scattering accompanying activity in nerve, *J. Cell. Comp. Physiol.* **40,** 199–219.

Carnay, L. D., and Barry, W. H. (1969). Turbidity, birefringence and fluorescence changes in skeletal muscle coincident with the action potential, *Science* **165,** 608–609.

Cohen, L. B. (1970). Light scattering changes during axon activity, *Permeability and Function of Biological Membranes,* L. Bolis, A. Katchalsky, R. D. Keynes, W. R. Loewenstein, and B. A. Pethica (ed.), New York, Elsevier, 318–325.

Cohen, L. B., Hille, B., and Keynes, R. D. (1969). Light scattering and birefringence changes during activity in the electric organ of *Electrophorus electricus, J. Physiol.* **203,** 489–509.

Cohen, L. B., Keynes, R. D., and Hille, B. (1968). Light scattering and birefringence changes during nerve activity, *Nature* **218,** 438–441.

Conti, F., Tasaki, I., and Wanke, E. (1971). Fluorescence signals in ANS-stained squid giant axons during voltage-clamp, *Biophysik* **8,** 58–70.

Foreman, J. C., Mongar, J. L., and Gomperts, B. D. (1973). Calcium ionophores and movement of calcium ions following the physiological stimulus to a secretory process, *Nature* **245,** 249–251.

Hille, B. (1970). Birefringence changes in active cell membranes, *Permeability and Function of Biological Membranes,* L. Bolis, A. Katchalsky, R. D. Keynes, W. R. Loewenstein, and B. A. Pethica (ed.), New York, Elsevier, 312–317.

Mueller, P., and Rudin, D. O. (1968a). Action potentials induced in bimolecular lipid membranes, *Nature, Lond.* **217,** 713–719.

Mueller, P., and Rudin, D. O. (1968b). Resting and action potentials in experimental bimolecular lipid membranes, *J. Theoret. Biol.* **18,** 222–258.

Narahashi, T., Albuquerque, E. X., and Deguchi, T. (1971). Effects of batrachotoxin on membrane potential and conductance of squid giant axons, *J. Gen. Physiol.* **58,** 54–70.

Narahashi, T., Moore, J. W., and Scott, W. R. (1964). Tetrodotoxin blockage of sodium conductance increase in lobster giant axons, *J. Gen. Physiol.* **47,** 965–974.

Ross, W. N., Salzberg, B. M., Cohen, L. B., and Davila, H. V. (1974). A large change in dye absorption during the action potential, *Biophysic. J.* **14,** 983–986.

Spyropoulos, C. S. (1972). Some observations on the electrical properties of biological membranes, *Membranes; Macroscopic Systems and Models* (vol. 1), G. Eisenman (ed.), New York, Dekker, 267–317.

Steinhardt, R. A., Epel, D., Carroll, E. J., and Yanagimachi, R. (1974). Is calcium ionophore a universal activator for unfertilized eggs? *Nature* **252,** 41–43.

Szabo, G., Eisenman, G., Laprade, R., Ciani, S. M., and Krasne, S. (1973). Experimentally observed effects of carriers on the electrical properties of bilayer membranes — equilibrium domain, *Membranes; Lipid Bilayers and Antibiotics* (vol. 2), G. Eisenman (ed.), New York, Dekker, 179–327.

Tasaki, I., Hallett, M., and Carbone, E. (1973). Further studies of nerve membranes labelled with fluorescent probes, *J. Memb. Biol.* **11,** 353–376.

Tasaki, I., Watanabe, A., and Hallett, M. (1972). Fluorescence of squid axon membrane labelled with hydrophobic probes, *J. Memb. Biol.* **8,** 109–132.

ANSWERS TO PROBLEMS

INDEX

ABOUT THE BOOK

The text of this book was set in Lino-
film Century Schoolbook by Ruttle,
Shaw & Wetherill, Inc. The editor was
Robert H. Warner, Jr. The designer
was Joseph Vesely, who supervised
production and make-up of page
mechanicals. Illustrations were drawn
at Vantage Art, Inc. Murray Printing
Company manufactured the book.